HUMAN FACTORS IN AVIATION OPERATIONS

Human Factors in Aviation Operations

Proceedings of the 21st Conference of the European
Association for Aviation Psychology (EAAP)
Volume 3

edited by
RAY FULLER
NEIL JOHNSTON
NICK McDONALD

AVEBURY
aviation

Published by
Avebury Aviation
Ashgate Publishing Limited
Gower House
Croft Road
Aldershot
Hants GU11 3HR
England

Ashgate Publishing Company
Old Post Road
Brookfield
Vermont 05036
USA

British Library Cataloguing in Publication Data

Human Factors in Aviation Operations:
Proceedings of the 21st Conference of the European
Association for Aviation Psychology (EAAP)
    I.Fuller, Ray
    629.130019
ISBN 0-291-39825-1

Library of Congress Catalog Card Number: 94-72810

Printed in Great Britain at the University Press, Cambridge

Contents

Part 2 ATC: Automated Systems

Part 3 ATC: Human Factors

Introduction

Ray Fuller, Neil Johnston and Nick McDonald,
Aerospace Psychology Research Group, Trinity College Dublin

Human factors are now universally recognised to be at the very heart of future developments in the maintenance and improvement of aviation safety. It is not surprising, therefore, that this subject represents an ever increasing proportion of the research agenda of aviation psychologists, that it is emerging as an essential element in undergraduate and postgraduate degree courses in applied psychology across the world, and that it is now formally incorporated as a required element in the training and assessment for pilot licensing.

This book is the third in a series of three, volume I entitled *Applications of Psychology to the Aviation System* and volume II being *Aviation Psychology: Training and Selection*. Taken together these three volumes provide a valuable and timely overview of the present state of aviation psychology, especially within Europe, and all three complement the previously published *Aviation Psychology in Practice*.

Collectively, these three volumes contain the proceedings of the twenty-first conference of the European Association for Aviation Psychology (EAAP). In recognition of the extensive political and structural changes currently taking place within Europe, WEAAP the former Western European Association for Aviation Psychology -was renamed EAAP at the conclusion of this conference.

This volume is structured in twelve parts, each of which addresses aspects of human factors in aviation psychology under the following broad headings:

- Aeronautical Decision Making
- ATC - Automated Systems
- ATC - Human Factors
- Critical Incident Stress Management
- Error Analysis

1

- Fear of Flying
- Hardware and Software Interface Design
- Aircraft Maintenance
- Physiological Factors
- Pilot Competence
- Situation Awareness
- Workload

The contents reflect in a comprehensive way a range of current human factors issues in aviation, including much ongoing work concerned with the effects of new technology on human performance and implications of new technology for the role of the human operator and for training. This applies in particular to the introduction of automated cockpits, flight management systems and datalink implementations in ATC. Reflecting the more cognitive thrust of the current *Zeitgeist,* much of this work involves the development and evaluation of models of human cognitive performance .

In a more traditional vein there are a number of studies reporting the effects on performance of stressors of various kinds, including emergency evacuations, total power failure in helicopters, excessive workload, mild hypoxic hypoxia and fear of flying. The research typically goes well beyond a descriptive analysis of the problem to the evaluation of management strategies, treatments (including those for post traumatic effects) and training interventions.

Research on the effectiveness, efficiency and safety of design characteristics of aviation hardware encompasses here topics such as integrated displays, datalink messages, ground proximity warning systems, head-up displays and colour coding in head-up displays. Related research starts from the other end of the machine-human system and explores the implications of human characteristics (such as cognitive operations and visual search strategies) for the design of other components in the system.

Lastly there is still much work being reported which is concerned with more fundamental aspects of human aviation psychology. Some of the examples in this volume include the analysis of factors which degrade flight monitoring, which affect perception of motion, and which affect situational awareness.

This brief introduction cannot do justice to the richness of the current research material, concepts, models and findings reported in this volume: the impact of developing technologies on airspace and aircraft management by ATC and flight crew, the problem of optimising the integration of the human component with technically driven automated and intelligent system components, and the impact on performance of "traditional" human factors such as stress and more recently the post-traumatic stress syndrome. Suffice it to say that these broad themes must recommend its contents to a wide range of applied psychologists, not only those in aviation, but also trainers and operational aviation personnel at all levels.

Ray Fuller, Neil Johnston and Nick McDonald

June 1994

Part 1
AERONAUTICAL
DECISION MAKING

1 Aeronautical decision making in general aviation: new intervention strategies

R.S. Jensen, G.P. Chubb, J. Adrion-Kochan, L.A. Kirkbride and J. Fisher
The Ohio State University, Columbus, Ohio

This is a discussion of a new approach to the development of intervention techniques for aeronautical decision making in general aviation under investigation at the Ohio State University. The study includes the development of models of the human decision process and applies these models to the general aviation domain using the expertise approach. Other aspects of the study include the development of personal minimums checklists for general aviation, and an examination of strategies for the dissemination of safety information to general aviation pilots.

Cognitive Modeling for General Aviation

Training strategies for the improvement of general aviation pilot judgment based on attitude models of human information processing have been developed and tested with considerable success in the USA, Canada, and Australia (Diehl, 1992). However, there is a growing belief that the usefulness of this approach may have reached a plateau and that it is time to consider a more comprehensive model as the basis for new judgment intervention strategies for general aviation.

The objective of this research is to develop new intervention strategies to improve aviator decision making using an expertise approach to cognitive task analysis. Traditional Instructional Systems Design (ISD) methods do not offer useful guidelines for complex cognitive tasks such as those involved in flying today's complex aircraft (Ryder and Redding, 1993). Traditional statistical methods in cognitive task analysis do not easily yield intervention strategies to correct problems that may be identified. The expertise approach is being used because its application can reduce duplication of effort in the development of intervention strategies.

The Expertise Approach

The study of expertise seeks to understand and account for what distinguishes outstanding individuals in a domain from less outstanding individuals in that domain, as well as from people in general. The approach is limited to outstanding behavior that can be attributed to relatively stable, learned characteristics of the relevant individuals. The classical expertise literature says that expertise is acquired over decades of experience and practice. The approach includes both a careful analysis of the attained performance and an analysis of how that performance was attained.

Building cognitive models using the expertise approach involves three steps: 1) Identifying representative tasks that capture the essence of superior performance in a specific domain, 2) Detailed analysis of the superior performance through several methods including verbal reports during performance of the tasks (or stimulus materials), 3) Efforts to account for the acquisition of the characteristics and cognitive structures found to mediate superior performances of experts.

The identification of a collection of tasks that can capture superior performance is often not easy. It may be helpful to identify only a small number of such tasks and test them on a larger population of experts and novices in a particular domain. Alternatively, one might identify a large number of tasks to be used on a small number of subjects.

The full range of methods of analysis in cognitive psychology can be applied in the examination of phenomena associated with a particular type of expertise including, performance analysis, expert-novice comparisons, and extensive studies of single subjects. One cannot, of course, directly observe the mediating steps in information processing but it is possible to specify hypotheses about relations between the internal processing steps and observable behavior. Think-aloud verbalization and retrospective verbal reports are the most common methods of assessing differences in mediating processes as functions of the subject's level of expertise. Subjects are asked to think aloud while carrying out a series of representative tasks.

Although the expertise literature says that aggregation of experience is the single most important factor in the acquisition of expertise (Chase and Simon, 1973), we believe that "meaningful" experiences are at least important if not

6

essential for learning to occur. For example, some tasks never seem to be learned even though they are performed repeatedly. Several studies have demonstrated that, at best, minor improvement is gained with years of experience in comparison with initial training (Ericsson & Smith, 1991). In aviation, pilots know that flight experience does not equate to expertise -- yet, hours of flight time continues to be the criterion for currency, ratings and certificates. We believe that other important mechanisms at work include: the teacher, instructional media, motivation, and internal adaptation processes.

Compared with statistical procedures, expertise procedures are not likely to yield more reliable results. However, by going through the expertise procedure, we are more likely to encounter and develop a better approach toward interventions than we would using the statistical procedures. This is because much of the expertise approach deals with the acquisition of expertise, not just the explanation of expertise. The statistical approach may retrospectively identify what happened and how often, but it does not suggest the causal factors or remedial actions. The expertise approach actively solicits such information and the results lead more quickly into intervention strategies.

The General Aviation Domain

The domain selected for this study of human expertise is a subset of general aviation pilots who fly the Cessna P-210, the Pressurized Beech Baron, and the Piper Malibu. These are considered complex, fairly high performance single- and twin-engined airplanes and are usually flown by a single pilot, often into the most complex airspace in almost all weather conditions. For these reasons, the flying task may be as difficult as any civilian flying task. To add to the complexity of the task, many of the pilots of these aircraft have other professions (doctors, lawyers, etc.) and do not fly very often and, therefore, may not be as proficient as pilots who make flying their profession.

Approach

The first step in our cognitive task analysis was an informal interview survey of pilots to determine a reasonable definition of expert vs novice performance in this domain. It should be noted that instead of the word "novice" we believe that "competent" is a better term to describe the type of person to which we are referring because these pilots have passed competency checks.

Distinguishing Qualities of the Expert General Aviation Pilot. The results of these interviews reveal that an expert in this domain is one who 1) possesses self-confidence in her flying skill and practices it constantly, 2) is highly motivated to learn all there is to know about this flight domain, 3) has superior ability to focus (or compartmentalize) one's attention on the flying task at hand,

4) possesses the mental discipline to change her focus of attention when new information suggests it is necessary, 5) is a keen observer of the flight environment, including location of other aircraft, terrain, navigation features, ATC clearances, weather phenomena, 6) carefully establishes a baseline for normal instrument indications, aircraft sounds, vibrations, and g-forces with respect to control action so that his or her threshold for slight variations is very small, 7) is continuously skeptical about "normal" aircraft functioning, 8) constantly makes contingency plans for those circumstances when things might go wrong, 9) possesses superior mental capacity for problem diagnosis, risk assessment, and problem resolution, 10) has excellent communication skills and who can readily adapt them to the audience and situation, and 11) knows her limitations, is motivated to avoid situations that might push her skill to those limits, and has the willpower to overcome the pressures of people around her to push the limits of her skill.

Conversely, a competent pilot in this domain is one who does not possess these superior qualities of the expert. The competent pilot may have as much flight time as the expert, has the stick and rudder skills sufficient to pass all necessary flight tests, and has the knowledge necessary to pass the written tests. Although he or she has taken the biennial flight review but that is as far as he or she takes aviation. Because motivation in these pilots is focused on matters outside of the cockpit, the extra skills, knowledge, mental models found in the expert have not been developed in the competent pilot.

In addition to these personal motivational factors, some other reasons why the competent pilot often does not become an expert include the organization (bureaucracy) for which he works, automatic systems in the aircraft, and creativity blocking standard operating procedures for the aircraft which he or she flys. The organization can block a person from becoming an expert by turning down his every request for changes that she has found from her experience are needed for safe efficient operations. Automatic systems can block a pilot from becoming an expert by forcing him to following symbols and control actions that are designed into the system without thought for many novel situations offered in aviation. Standard operating procedures and regulations can block a person from becoming an expert by forcing her to comply with procedures rather than do what she knows to be safer and more effective (Fahlgren and Hagdahl, 1990).

An examination of the qualities that differentiate the expert from the competent general aviation pilot reveals that much of the task involves cognitive activity. "Stick and rudder" skills play but a small role in differentiating the expert from the competent. If one only considers the tasks necessary to be a competent general aviation pilot, the tasks might be more heavily weighted on the perceptual-motor skills. However, if one wishes to be an expert in this domain, one must excel in cognitive skills.

Distinguishing Domain-Specific Tasks. The next task involves the design of tasks or scenarios that distinguish the expert from the novice in this general aviation domain. In doing so one must also be careful to develop tasks that can be done in a laboratory simulation. We believe that the following tasks fit these criteria: 1) wind shear detection and recovery at a low altitude, 2) subtle degradation of an engine instrument, 3) gradual build up of ice on the tail surfaces, and 4) an airport closure during a hold awaiting approach clearance. These are among the tasks currently being developed for use in our testing program to distinguish the expert from the novice.

Completing the Cognitive Task Analysis. Some of the expertise methods that we are using include structured interviews with pilots alone where domain specific scenarios are resolved, working out these same scenarios in groups, and resolving similar scenarios in simulator exercises. Pilot behaviors are recorded on video tape for later analysis. Retrospective verbalization of the problem resolutions are also being used in debriefing following the simulator exercises. Verbal data recorded on video tape will be analyzed through coding into the categories that identify the expert from the novice in this domain. From these data will emerge our model of how the expert general aviation pilot differs from the novice together with ideas on how one can make experts out of the novices in this domain.

Intervention Strategies. The record shows that it is in the area of cognitive skills where we fail most often. Deliberate teaching of these judgment skills in general aviation (where the foundation for a flying career is established) is almost non-existent. A strong emphasis is being placed on the development of some of these skills in the airline environment but, at that stage, it may be too late for some to change habit and thinking patterns learned in general aviation.

Our approach in this research is to develop new cognitive models of the cognitive processes applied to this general aviation domain and to use these models to develop new approaches to modifying pilot decision making processes. The following is a discussion of two other aspects of this study that are aimed at improving interventions into decision making in this domain.

Personal Minimums Checklist Study

Another task in this program is to develop a checklist for pilots to use to aid general aviation pilots in their evaluation of various (flight) safety related factors in a structured manner for go/no go decisions using a relative risk assessment technique. Preliminary steps include: (a) generating a checklist of factors contributing to such decisions, (b) creating prototype guidelines for generating personalized minimum operational standards based on this checklist, and (c) providing a detailed plan for validating the checklist and guidelines for general aviation.

The Marketing of Safety Information to GA Pilots

Finally, the FAA, along with other organizations, conducts a variety of programs aimed at improving aviation safety among the pilot population. Because of the self-selection of attenders at these seminars, the representativeness of the audience may be questioned. Do the attenders represent that portion of the pilot population at greatest risk of an accident, or do these seminars "preach to the choir?" If the latter is the case, then how may the FAA better disseminate its information of aviation safety to those most in need of its benefits? The objective of this task is to identify the needs and requirements for disseminating safety-related information to pilots and to develop a cost effective/training effective strategy for reaching all segments of the pilot population.

Acknowledgements

The work here reported is part of a new human factors research support program sponsored by the Federal Aviation Administration in the USA. The Contract Monitor is Dr. David Hunter. The ideas and data reported are those of the authors and do not necessarily reflect those of the sponsoring organization.

References

Chase, W.G. & Simon, H.A. (1973). Perception in Chess. *Cognitive Psychology, 4*, 55-81.

Diehl, A. (1992, January). Does cockpit management training reduce aircrew error? *ISASI Forum, 24*(4).

Ericsson, K.A. & Smith, J. (Eds) (1991). *Toward a general theory of expertise: Prospects and limits.* Cambridge: Cambridge University Press.

Fahlgren, G. and Hagdahl, R. (1990). Complacency. *Proceedings of the 43rd Annual International Air Safety Seminar,* Rome, Italy: Flight Safety Foundation.

Ryder, J.M. & Redding, R.E. (1993). Integrating cognitive task analysis into instructional systems development. *Educational Technology Research & Development, 41*(2), 75-96.

2 Aeronautical decision making: historical results and a new paradigm

David R. Hunter, PhD
Federal Aviation Administration, Washington DC

Pilot error has long been implicated as a predominant cause of aircraft accidents. Nagel (1988) notes that the cockpit crew was cited as a causal factor in 67 percent of the air carrier accidents among the worldwide jet fleet from 1959 to 1983. Similarly, data from various studies estimate that pilot error accounts for 65 to 85 percent of general aviation accidents. In a review of the general aviation accidents between 1970 and 1974, Jensen & Benel (1977) found that poor flying judgment was involved in 35 percent of all non-fatal and 52 percent of all fatal accidents. Jensen & Benel suggested that judgment could be taught and evaluated and proposed a two component definition of flying judgment which included: "The ability to search for and establish the relevance of all available information regarding a situation...[and]...The motivation to choose and authoritatively execute a suitable course of action..." (Jensen & Benel, 1977, p 34)

Based predominately on the second part of that definition, an intervention strategy was developed by Berlin et al. (1982) which attempted to improve decision making by making pilots more aware of how decisions are made, the affective or attitudinal components of decision making, and procedures which may be followed to improve the quality of decisions. Berlin et al. categorized poor judgment as involving one or a combination of three subject areas: pilot, aircraft, or environment. They further described the pilot's action associated with an accident as: Do-No Do; Under Do - Over Do; and, Early Do - Late Do.

11

In addition, they described the poor judgment behavior chain, three mental processes of safe flight, and the five hazardous thought patterns (anti-authority, impulsivity, invulnerability, macho, and external control). The presence or influence of these patterns is suggested as the cause of the "irrational pilot judgment" described by Jensen & Benel (1977).

In an evaluation of judgment training conducted by Berlin, et al., the experimental group (which received the judgment training) was significantly better than the control group in decision making as measured by the number of incidences of hazardous behavior in a check flight. Within the several limitations on generalizability noted by Berlin, et al, these results showed that the training could improve the quality of decision making. Berlin, et al, also offered suggestions for additional evaluation approaches, including assessment of the duration of effectiveness of the training, involvement in accidents, and other behavioral indices; unfortunately these additional evaluations have not been accomplished.

This approach at intervention, which is often referred to as hazardous thoughts training, was subsequently used by the Federal Aviation Administration in the development of several products aimed at the general pilot population and specific subgroups. It has also being used, to some degree, worldwide in a commercial training program developed by Bell Helicopter Company for its customer fleet (Fox, 1991) and in programs conducted by civil aviation authorities in Canada and Australia.

Table 1
Judgment Training Experimental Evaluations

Citation	Subjects	Results[1]
Telfer (1989)	Australian private pilots	8%
Buch & Diehl (1983)	Canadian private pilots	9%
Diehl & Lester (1987)	United States private pilots	10%
Berlin, et al. (1982)	United States student pilots	17%
Buch & Diehl (1984)	Canadian civilian cadets	40%
Connolly, Blackwell, & Lester (1989)	United States instrument students	46%

NOTE 1: Results are stated as percentage of errors or hazardous behaviors committed by the treatment group compared to a control group. Adapted from Diehl (1991).

The effectiveness of this approach has been evaluated in several studies in the United States and elsewhere. As shown in Table 1, these studies have achieved mixed results, with effects ranging from an 8 percent reduction in hazardous behaviors among private pilots to a 46 percent reduction among instrument students attending training in a university setting. In addition, a recent study by Edens (1991) found "None of the hazardous thought patterns assessed were significantly related to either attention/perception or judgement/decision errors." (p. 98)

Although at present the hazardous thoughts training may represent the best available approach to pilot decision making improvement, the results suggest that the magnitude of the effect varies substantially among groups and settings, with the best results apparently being obtained in highly structured training settings. The suitability of this approach is therefore questionable for pilots outside those settings. In addition, as Edens (1991) and Telfer (1989) note, there are substantial methodological issues which have not been addressed in the studies to date and, "It would be quite inappropriate to conclude, simply, that the efficacy [of this approach] has been demonstrated." (Telfer, 1989, p 170).

Among the issues that have not been addressed is the mechanism of effect of this approach. Hazardous thoughts training is aimed at altering relatively enduring and extensive personality attributes which may characterize an individual's functioning well beyond the cockpit. Clinical experience has shown us that such attributes are not easily modified, even in much more extensive intervention settings; hence, it is questionable whether the brief exposure to these concepts given during flight training is likely to have an enduring effect. These methodological and validity issues, in combination with more pragmatic concerns regarding the marketability of the interventions derived from this model, have led to a reexaminination of our approach to understanding pilot decision making. Our focus has shifted from addressing pilot decision making in terms of attitudes and personality constructs to developing an understanding of the pilot as an acquirer and processor of data, in accordance with current theories of human information processing (cf Barnett, 1989; Wickens, 1984).

Multiple models of the human as an information processor exist and analyses are possible on a number of levels of complexity, from simple linear models which examine only differential attribute weights to much more highly structured models based upon theoretical concepts of cognitive structures and processes (Kahneman & Tversky, 1973; Klein, 1991; Ericsson, 1991). The choice of a particular model to use is based not only on its veridicial properties but also on its practical utility. Simplistic models may not be true representations of underlying cognitive processes, but for the purposes of prediction of behavior they may be acceptable approximations. If they also possess properties that make them amenable to application in marketable intervention programs, then they may prove as worthwhile as more valid, but complex, models which do not lend themselves to ready marketability. In recognition of this continuum, the strategy which we have adopted in our emerging research program is to use a two-pronged approach. In parallel efforts being conducted by independent organizations, we are investigating pilot decision making using models that range from a simple, atheoretical linear model to models which represent highly developed theories of cognitive structure and processing.

In a study reported at the First International Symposium on Aviation Psychology, Flathers, Giffin, & Rockwell (1981) used a linear modeling technique called conjoint analysis to assess the attribute worth functions of pilots. This process examines the values or utilities which pilots apply to the decisional attributes in arriving at a final decision. Flathers et al. used a paper-and-pencil simulation of a flight during which a minor emergency (alternator failure) occurred forcing diversion to an alternate airport. The attributes manipulated were: ATC services at the airport; weather at the airport; time to fly to the airport; and, best instrument approach facilities. Each of these attributes was varied over two levels. Subjects performed a card sorting exercise in which they rank ordered the airports from most to least preferable, given the situation they were in.

From these data, Flathers et al, were able to construct linear models of the subjects decision making process which accounted for over 90% of the response variation, confirming the appropriateness of the additive linear model. We will use a similar linear modeling technique to investigate the pilot decision processes related to flight into adverse weather conditions. Encounters with adverse weather consistently comprise the largest single cause of fatal accidents. In a study of accidents between 1975 and 1986, the National Transportation Safety Board (NTSB) found that "accidents involving visual flight rule flight into instrument meteorological conditions accounted for 4 percent of all general aviation accidents but produced 19 percent of the resulting fatalities." (NTSB, 1989, pg 1) The pilots involved in these accidents were predominately private pilots with less than 500 total flight hours and were not instrument rated. Pilots like these make up a large portion of the non-commercial pilots flying in the United States and represent a significant challenge to the development of suitable interventions and marketing strategies.

A substantially different scenario is represented in the problem to be addressed using the more theoretically based cognitive processing model. Pilots flying high-performance general aviation aircraft (such as the pressurized Cessna 210 and the Malibu) at altitudes from 12,000 to 24,000 feet encounter a substantially different flying environment and are typically more highly experienced with higher qualifications than private pilots flying low-performance, single-engine aircraft into adverse weather. Yet in this environment too, pilots make decisions which lead to accidents -- typically involving incursions into icing and other severe weather conditions or exceeding the operating limitations of the aircraft. A cognitive processes approach has been chosen for analysis in this case because the highly complex aircraft and systems involved do not seem amenable to description in terms suitable for linear models. I will not review this approach in detail, as it will be described by Professor Jensen in his presentation. The interested reader is therefore referred to Jensen's article in these proceedings for further information.

In both these scenarios our intent will be to develop an understanding of the decision process sufficient to construct effective and efficient interventions. To do so, we must examine the situational specificity of the decision parameters, their reliability, the degree to which they may be modified and under what circumstances that may take place. Comparisons of the decision parameters of novice, intermediate, and expert pilot groups may also provide insights into the development of decision making expertise which could be used to guide intervention development, and these will certainly be conducted.

Although the initial stages of our research effort will involve basic studies of decision making by pilots, the goal is not simply to produce new knowledge, but to produce effective products based upon that knowledge. Thus, while these efforts are underway to develop the models of decision making necessary for understanding the processes, another effort is underway to examine the characteristics of the pilot population from a marketing standpoint in order to better understand how we may impact behavior. This effort will develop the marketing strategies necessary to effectively and efficiently take transform the technologies developed by our research into a useable and acceptable product.

The culmination of these converging efforts will produce a variety of interventions tailored to reach the pilots most at-risk for accidents in a effective and efficient way. We believe that this program will significantly reduce the incidence of accidents attributable to the major cause of fatal accidents -- poor decision making -- and will save many lives in the years to come.

References

Barnett, B. J. (1989). Information processing components and knowledge representations: An individual differences approach to modeling pilot judgement. *Proceedings of the Human Factors Society 33rd Annual Meeting*, 878-882.

Berlin, J. I., Gruber, E. V., Holmes, C. W., Jensen, P, K., Lau, J. R., Mills, J. W., & O'Kane, J. M. (1982a). *Pilot Judgement Training and Evaluation - Volume I.* (DOT/FAA/CT-82/56-I, II, III). Washington, DC: Federal Aviation Administration.

Buch, G., & Diehl, A. (1983) Pilot judgment training manual validation. Unpublished report. Ontario: Transport Canada. As cited in Diehl, A.

(1991). The effectiveness of training programs for preventing aircrew error. In R. S. Jensen (Ed.), *Proceedings of the Sixth International Symposium on Aviation Psychology.* Columbus, OH: Ohio State University.

Buch, G., & Diehl, A. (1984). An investigation of the effectiveness of pilot judgement training. *Human Factors, 26,* 557-564.

Connolly, T. J., Blackwell, B. B., & Lester, L. F. (1989). A simulator-based approach to training in aeronautical decision making. *Aviation, Space and Environmental Medicine, 60,* 50-52.

Diehl, A. E. (1991). The effectiveness of training programs for preventing aircrew error. In R. S. Jensen (Ed.), *Proceedings of the Sixth International Symposium on Aviation Psychology.* Columbus, OH: Ohio State University.

Diehl, A. E., & Lester, L. F. (1987). *Private Pilot Judgement Training in Flight School Settings.* (DOT/FAA/AM-87/6). Washington, DC: Federal Aviation Administration.

Edens, E. S. (1991). *Individual Differences Underlying Pilot Cockpit Error.* (DOT/FAA/RD-91/13) Washington, DC: Federal Aviation Administration.

Einhorn, H.J., & Hogarth, R.M. (1981). Behavioral decision theory: Processes of judgment and choice. *Annual Review of Psychology, 32,* 53-88.

Ericsson, K. A. (1991). *Toward a General Theory of Expertise.* New York: Cambridge University Press.

Flathers, G.W., Giffin, W.C., & Rockwell, T.H. (1981). A study of decision-making behavior of aircraft pilots deviating from a planned flight. In R. S. Jensen (Ed.), *Proceedings of the Fourth International Symposium on Aviation Psychology.* Columbus, OH: Ohio State University.

Fox, R. G. (1991). Measuring safety in single- and twin-engine helicopters. *Flight Safety Digest,* July 1991.

Jensen, R. S., & Benel, R. A. (1977). *Judgement Evaluation and Instruction in Civil Pilot Training.* (FAA-RD-78-24). Washington, DC: Federal Aviation Administration.

Kahneman, D., & Tversky, A. (1973). On the psychology of prediction. *Psychological Review, 80,* 237-251.

Klein, G. A., & Zsambok, C. E. (1991). Models of skilled decision making. *Proceedings of the 35th Annual Human Factors Society Meeting.* San Francisco:

Nagel, D. (1988). Human error in aviation operations. In E. Weiner & D. Nagel (Eds.), *Human Factors in Aviation.* San Diego: Academic Press.

NTSB (1989). *General Aviation Accidents Involving Visual Flight Rules Flight into Instrument Meteorological Conditions.* (NTSB/SR-89/01). Washington, DC: National Transportation Safety Board.

Telfer, R. (1989) Pilot decision making and judgement. In R. S. Jensen (Ed.) *Aviation Psychology.* Brookfield, VT: Gower Technical.

Wickens, C. D. (1984). *Engineering Psychology and Human Performance.* Columbus, OH: Charles E. Merrill Publishing Co.

16

3 FOR-DEC: a prescriptive model for aeronautical decision making

Hans-Jürgen Hörmann
Deutsche Forschuungsanstalt fuer Luft- und Raumfahrt (DLR)
Institute of Aerospace Medicine, Department of Aviation and Space
Psychology, Hamburg, Germany

Summary

The paper describes the development of a model, which is used as a framework for the training of judgement and decision making in CRM courses for airline pilots. The acronym FOR-DEC stands for six different phases of the decision making process: Facts, Options, Risks & Benefits, Decision, Execution and Check. The advantages of such a simple prescriptive model are that it can easily be remembered and that it helps to structure judgemental and decision making processes in the cockpit. It therefore counteracts certain cognitive mechanisms that can adversely affect the quality of aeronautical decisions in terms of flight safety.

Introduction

This paper describes a part of the Lufthansa CRM-training which was developed in a working group consisting of human factors specialists of the German Aerospace Research Establishment (DLR) and of airline pilots from Lufthansa and Condor. As mentioned in the paper by Maschke et al. (1994), this training programme covers primarily three broad subjects: communication, teamwork, and judgement & decision making. The model described in this paper is included in the units on judgement and decision making.

Prescriptive models for aeronautical decision-making (ADM) are defined in the literature as "quick and easy" heuristic methods to structure the cognitive process of decision making (O'Hare, 1992). They are usually suggesting a sequence of steps which should be followed to improve decision quality. These steps are often summarized by acronyms such as "DECIDE" (Detect, Estimate, Choose, Identify, Do, Evaluate) (ref. Clarke, 1986) or "PASS" (Problem, Acquire, Survey, Select) (ref. Maher, 1989). These models are neither based on a certain theory nor do they claim to lead always to optimal decisions.

The model shown here is based on the characteristics of the human decision maker in complex, dynamic situations, the typical working environment of pilots. These situations are associated in many cases with time pressure, continually changing conditions, distractions, and incomplete information (Klein & Klinger, 1991). Additionally, decisions are often not reversible and errors can cause severe personal consequences. Various laboratory studies have shown that the quality of human decisions generally decrease substantially under such conditions (see Boff & Lincoln, 1988). The human decision maker tends for example

- to formulate a hypothesis early and then tries to confirm rather than to test it
- to be too conservative in recognizing changes in problem conditions
- to adopt the first solution developed
- to be reluctant to change an erroneous commitment in light of new evidence
- to ignore ambiguous or partial data
- to overestimate the probability of favourable outcomes, and to underestimate the probability of unfavourable outcomes

These cognitive mechanisms are partially due to limitations of the human information processing system itself. On the other side they are related to attitudes and habits of the individual and hence can be influenced by appropriate training and design.

Considering these characteristics a prescriptive model for ADM was developed which had

- to take into account the complex, dynamic environment of the cockpit work
- to be applicable not only to abnormal situations but also to

different classes of decisions during normal operation (instant decisions, procedural decisions, analytic decisions)
- to separate the phases of collecting and evaluating possible solutions
- to provide a handy tool, being easily remembered and inviting itself for application at the workplace.

FOR-DEC model for individual decision making

FOR-DEC is a made-up word to symbolize six different phases of the decision making process: Facts, Options, Risks & Benefits, Decision, Execution, Check. Each phase is connected to a guiding question which should help to focus one's attention on a sequence of essential steps for effective decision making. During the interaction between the acting pilots and the operational environment, facts are the trigger for a decision making process. A need for decision is recognized. After one or more cycles the process stops, when a check of executed decisions suggests that the desired results have been reached and the initial task is solved. If uncontrolled dynamics of the situation produce unexpected changes, the decision task itself may be redefined during one or more FOR-DEC cycles. "Sub-FOR-DECs" are also conceivable during each phase of the process. For example, it might be necessary to collect additional data, in order to assess a certain risk potential for a given option. The single phases of the FOR-DEC model as shown in Figure 1 are described below.

Facts: What is actually going on here? After the decision need is recognized, facts are collected to assess the situation and to figure out task priorities and goal constraints. The situation analysis should be cross-checked by reference to independent sources of information.

Options: What are the choices we've got? Following a verified situation analysis optional responses are generated. For the moment the strict evaluation of the options has to be postponed, because it could narrow the scope of solutions too early. If the situation analysis indicates a procedural or instant decision, which has to be handled by a standard operating procedure this predefined procedure is activated.

Risks & Benefits: What is there to be said for and against the application of the different options? Procedures or options are checked for applicability. Expected benefits and potential risks are assessed and compared for the different options. Because of the limited predictability of future events, the decision maker has to be aware of uncertainties (e.g. visibility at Minimum Descent Altitude).

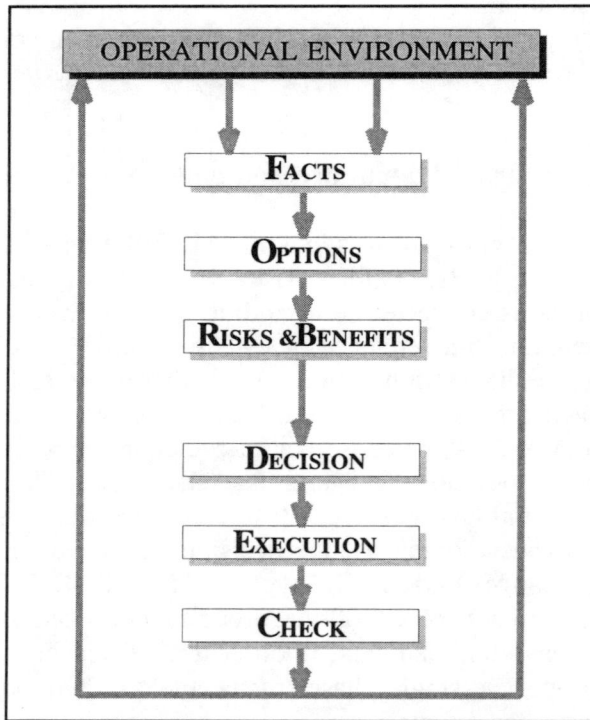

Figure 1   FOR-DEC model for aeronautical decision making

Decision: So, what shall we do after all? The most appropriate option in terms of minimum risks and maximum benefits is selected, provided that the situation analysis is still valid. To take possible uncertainties into account a back-up option should also be chosen.

Execution: Who shall do what, when, and how? The selected option is planned and executed in a coordinated manner using the available resources (including the SOPs) effectively.

Check: Is everything still allright? Actions and outcomes are monitored and compared with expected effects. The whole procedure is reviewed and updated for any unforeseen, overlooked, or new developments.

20

The acronym FOR-DEC offers some possible associations to the process of decision making, such as "FOR-DEC is a model *for decisions*". The dash in the middle symbolizes that before an adequate decision can be found, three preceding cognitive phases have to be passed. Of course the message of the model is neither to postpone a terminal decision too long nor to neglect other duties such as flying the aircraft first. It is intended to offer a guideline for how to cope with decision tasks with a more structured technique.

FOR-DEC model for crew decision making

As shown in Figure 2 the FOR-DEC model can be extended to crew decision making also. Each phase of the described process is initially passed independently by each individual crew member. The individual thoughts are then connected to and shared with the involved crew.

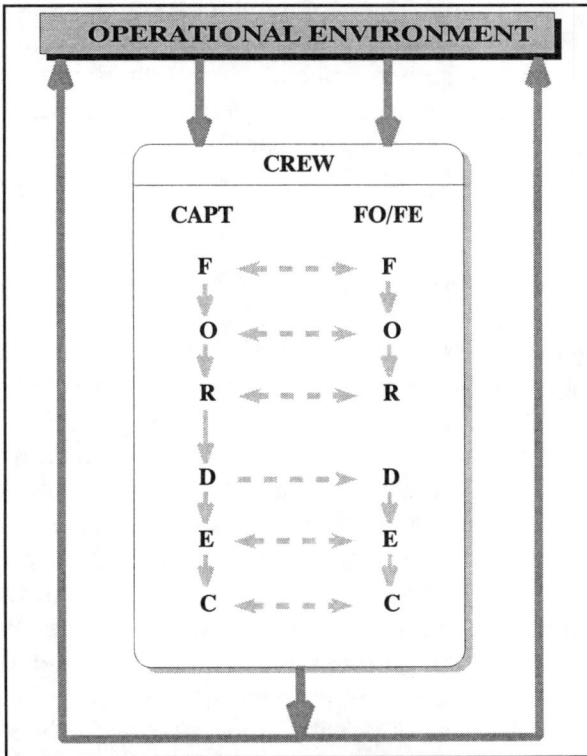

Figure 2   FOR-DEC model for crew decision making

21

To make the crew aware of the decision task at hand, facts should be announced and crosschecked. Options can be suggested and refined by all crew members. Since different utility aspects of possible solutions are often incompatible to each other these conflicts have to be resolved during the evaluation phase of risks and benefits. This consultation between the crew members during the first three phases should lead to a common mental model of the situation based on relevant facts rather than on opinions. If aggreement is reached at this point, the choice between alternative solutions is a direct consequence of the preceding steps. Normally the final decision is made by the pilot-in-command and communicated to the other crew members. During execution the planned procedure is briefed and tasks are assigned according to the manual or otherwise according to an adequate workload distribution. When the selected option is executed, actions are monitored and outcomes announced. Finally the crew should spend a moment for critique to review the applied procedure.

Provided that teamwork and communication between the crew members are intact, crew decisions are superior to those of the individual pilot in most cases. Two or more crew members can seek more information, can find more options, and are better in estimating possible risks and benefits than a single pilot. Furthermore tasks can be shared and a helpful feedback from the crew members during the Check-phase can be a valuable learning experience.

Training of decision making within a CRM-course

Several units of the DLR/Lufthansa CRM-course can be related directly to the FOR-DEC model. For example Situation Awareness can be related to Facts and Check. Attitudes & Risk Assessment to Risks & Benefits, Time Management & Workload Control to Execution. The main methods in these units are case studies, problem scenarios, crew problem solving games. Dynamic problems and creative exercises are used to practice the whole process of effective judgement and decision making in a team situation. In these team exercises we found that the FOR-DEC model also helps to structure communication and group interaction processes. While solving a complex problem in working-groups the pilots often used statements on a meta-level like, "let's have a look on what facts we have got, first", "have we considered the full scope of possible options?", or "let's go back and see, whether there is anything, we have overlooked". How it can be achieved that training with the FOR-DEC model has some positive transfer to real-life behavior outside the classroom is discussed by Schiewe in his paper on the "behavioral business card".

References

Boff, K.R. & Lincoln, J.E. (1988), *Engineering data compendium: Human perception and performance*, AAMRL, Wright-Patterson AFB, Ohio.

Clarke, R. (1986), 'A new approach to training pilots in aeronautical decision making', Frederick, M.D.: *AOPA Air Safety Foundation.*

Klein, G. & Klinger, D. (1991), 'Naturalistic decision making', *CSERIAC Gateway*, 2, 1-4.

Maher, J.W. (1989), 'Beyond CRM to decisional heuristics', *Proceedings of the Fifth International Symposium on Aviation Psychology*, Ohio State University, Columbus, 439-444.

Maschke, P., Goeters, K.-M., Hörmann, H.-J. & Schiewe, A. (1994), 'The development of the DLR/Lufthansa Crew Resource Management Training', *Paper presented at the 21st WEAAP conference in Dublin*, March 28-31.

O'Hare, D. (1992), 'The "artful" decision maker: A framework model for aeronautical decision making', *The International Journal of Aviation Psychology*, 2, 175-191.

Schiewe, A. (1994), CRM 'training and transfer: The "behavioral business card" as an example for the transition of plans into actual behaviour', *Paper presented at the 21st WEAAP conference in Dublin*, March 28-31.

23

4 Pilot decision making training: a Canadian application

James P. Stewart
Director General, System Safety, Transport Canada

What we accomplish in the area of human factors development and training results from the efforts of a number of people working together. This paper is no exception. I want to thank Mr. Norm LeBlanc, Mr. Bob Merrick, Ms. Susan Wood and Mr. Mike Doiron of the System Safety Directorate of Transport Canada for their contribution.

Background

With the introduction of the B-767, B-757 and, later, the A-320, aviation technology made an impressive leap forward. Hardware and software applications followed years of reducing aviation losses through more stringent airworthiness requirements and improved technology in aircraft design and construction. In 1993, for example, Air Canada experienced two in-flight engine shutdowns on their 767 fleet for every 100,000 engine hours. Their A-320 fleet experienced one in-flight shutdown per 100,000 engine hours[1]. However, progress on one of the most important components of our aviation system - **the liveware,** has not kept pace.

After years of study and discussion, human factors remains the last frontier in which major reductions in accidental loss may be achieved. Solutions to human performance limitations, to a great extent, remain elusive even though those performance limitations continue to dominate accident statistics.

Airmanship

Human limitations have been recognized for longer than I've been involved in aviation. We used to call it "airmanship". I can remember sitting in flight schools in Moose Jaw, Saskatchewan, and Cold Lake, Alberta, while instructors tried valiantly to teach me "airmanship". They always seemed a little surprised when I later went flying and occasionally proved that I still had a lot to learn.

The reality was that we only learned airmanship, or what passed for airmanship, after a number of years in the air[2]. Some learned "airmanship" quicker than others and those who did not occasionally ended up in smoking holes in the ground. If we were lucky, they only killed themselves.

But, we have not been without our successes and that is what I want to speak about to you today - a Canadian success story in responding to the human performance limitations of the one person who influences aviation events 100% of the time - **the pilot.**

Pilot Decision Making

Canadian statistics from 1976-1983 indicated that the human element was causal in about 84% of all civil aviation accidents in Canada[3]. In the United States, the National Transportation Safety Board arrived at the same conclusion. Worldwide, the ratio of human performance factors was about the same[4]. Within that context, judgement factors continue to dominate[5].

In 1982 Transport Canada and the Federal Aviation Administration signed an agreement to co-operate and share research results with a view to improving the pilot decision making process[6]. The joint research project spanned three years with major conferences being held at the end of each year to exchange information and findings. The final conference was held in Ottawa in December, 1986. Through these conferences, scientists, government officials and operational specialists hammered out a program to translate laboratory research into a specific action plan to address the need for implementing pilot decision making programs[7].

In December, 1986, **Pilot Decision Making** was born.

The Safe Program

In 1983, we developed an analytical tool to categorize and record factors which had been identified through investigations into aviation system failures (see Volume 1 for a discussion of the SAFE Program).

Judgement is the number one cause of accidents. Even though a major part of any pilot's life is spent observing conditions, detecting change and selecting alternatives, we seldom spend time on the teaching of effective decision making during basic training. Those of us who like to think we understand are still occasionally surprised when a pilot shows us, at some stage, that he or she still has a lot to learn. We have had numerous examples to remind us that we, as an industry, were doing something wrong. The Florida Everglades accident, the Stapleton accident in Denver, Palm Air in Washington, and the biggest of them all - Tenerife, all came down to faulty decision making of some sort.

Interestingly, pilots themselves supported the primacy of **Judgement** failure during a survey conducted by the Transportation Safety Board of Canada in 1991. Respondents attributed various cause factors to occurrences they had experienced. They identified **Judgement** as a factor in 59 % of accidents and 46% of incidents in which they were involved[8].

The Challenge Then and Now

Even with striking examples of major loss, early scepticism about our new direction was not easily overcome. We were working with what many in the aviation industry considered to be a "soft science". We were challenged to "prove your program is effective". We not only had to persuade people of the need for PDM, but had to show them some positive results fairly quickly if the program was to survive[9]. Finally, people perceived that we were trying to change the attitudes and practices of a professional group who were proud of their ability to cope and who had been trained to believe in that ability as a way of surviving. Many did not see the need for change.

What we were really teaching was an understanding of attitudes, how they affected a pilot's decision making and how they affected their interaction with their environment. We discussed the three components of Attitude - **Cognitive, Affective and Behavioural,** and dealt with the pilot and his environment within that context.

In the final analysis we far exceeded our expectations. Our original program was developed to deal with problems in the private pilot community. Over the years, we experienced a substantial level of commercial pilot "walk-ins". This caused us problems in developing specific programs to meet their needs. Our original three modules were expanded and we now have nine. Looking back, it seems incredible that we were concerned with survival and not concerned with how we were going to respond to demand for the program.

Initial Introduction

We started with three basic PDM modules to introduce aspects of PDM and we still use those modules today.

The first looks at the meaning of PDM, followed by an **Attitudes Inventory.** The Inventory has a total of ten scenarios that contain an explanation of the situation and then five choices. These five choices are the reasons why a particular action done in the explanation was carried out. Responses are scored and designed to assist pilots in determining their attributes as a decision maker. Upon completion they are shown how to score the inventory and prepare their personal profile. Results remain confidential. We then conduct class discussion to identify and understand each of the five hazardous attitudes towards risk taking.

Hazardous Thoughts	Antidote
Anti-Authority	Follow the Rules!
Impulsivity	Slow Down - Think First!
Invulnerability	It **could** happen to me!
Macho	Macho is Foolish!
Resignation.	I **can** make a difference!

The second module demonstrates a systematic method of making decisions and provides a new set of skills to be applied to all aspects of flying. The basis for this module is the DECIDE model originally developed by Mr. Ludwig Benner of Events Analysis Inc., to teach fire fighters in the United States.

Detect change
Estimate significance of change
Choose outcome objectives
Identify plausible action options
Do best option
Evaluate progress

Module three looks at managing stress and discusses causes and effects of stress.

Teaching Methods

We do not try to teach psychology. We do try to instill an understanding of decision making. We encourage behaviour modification by presenting examples and case studies. Operational safety officers teach PDM according to methods developed by knowledgeable psychologists and pedagogues[10]. "We present a systematic method of making decisions and provide pilots with new skills and knowledge which can be applied to all aspects of flying.

PDM workshops are organized in two hour sessions, preferably for an audience having common qualifications. Scenarios and case studies of actual occurrences relevant to the audience's particular operation serve as teaching/discussion tools to make the point(s) and help modify behaviour. Quality assurance teams monitor the workshops to ensure a national standard is achieved and maintained.

A major goal of the PDM program was achieved on May 30, 1991 when PDM training became mandatory in Canada and was introduced as part of the ab initio private and commercial pilot training requirements. No longer are we faced with changing inappropriate attitudes. We now develop positive attitudes from the beginning.

Present and Future Objectives

On January 1, 1990, our Program Development Committee proposed a major expansion of the PDM program. They proposed increasing the number of modules and developing new products to underline what had already been learned. All of these recommendations were accepted.

We developed additional modules under the PDM umbrella which have now moved out on their own and form an integral part of our overall safety awareness strategy. These include Risk Management Seminars, Human Factors in Communications and Safety Awareness for Flight Instructors. We also have PDM modules for different phases of flight (Pre-flight Planning, Enroute and Approach and Landing). We have just started work on a major safety awareness effort directed towards managers and we plan to launch that effort with a free newsletter to all company managers in Canada. And, lest you think we forgot, we are also discussing a separate course for aviation regulators.

In February, 1991, not satisfied with their first challenge, our Program Development Committee proposed adapting PDM to air traffic controllers, flight service specialists and flight attendants. At the same time, our Transportation Safety Board recommended that air traffic controllers receive decision making training similar to what we provided pilots. Everybody was jumping on the bandwagon. All of these programs are in preparation.

Conclusion

Safety awareness training (and PDM, in particular) has the potential to reduce risk in aviation and prevent loss. The results of PDM and similar programs have been reflected in a decline in pilot judgement factors as shown through our SAFE analysis. We hope to expand on that success by ensuring that more pilots and more aviation specialists receive such training.

We have learned that human factors problem solving and human factor programs are not easy to sell. It is difficult to persuade someone to spend money on a program whose effectiveness is hard to quantify. But, we have also learned that there is a substantial need for these programs and that the need is growing constantly as the complexity of aviation increases and the cost of accidental loss escalates.

Scientists, government officials, operational specialists and designers must continue to work together to ensure the dramatic advances in technology do not continue to exceed the human's ability to cope.

Bibliography

Buch, Dr. Georgette, *An Investigation of the Effectiveness of Pilot Judgement Training.* Human Factors, 1984, **26**(5),557-564

Caesar, Heino, *Have Due Regard for 50.000 Year Old, Software.* published in the Winter Edition of British Airways, Flight Deck magazine.

Clarke, Richard, *A New Approach to Training Pilots in Aeronautical Decision Making.*

Wallace, Garth, *PDM*. Canadian Aviation, February, 1987

Summary Proceedings of an International Industry/Government Workshop, Frederick, Maryland, March 14-16, 1985 (Final Report 1985), prepared by the Flight Safety Foundation, Arlington, Virginia

Pilot Judgement Training Validation Experiment done under the Joint Research project Agreement between Transport Canada and the Federal Aviation Administration, August, 1982.

Pilot Decision Making for General Aviation Pilots, Transport Canada

Commercial Pilot Survey (1991) Levels III to VI Air Carrier Operations, Conducted by the Accident Prevention Branch, Transportation Safety Board of Canada.

Notes

1   Information provided by the Analysis and Research Branch, Transport Canada.

2   Richard Clarke states, in his paper, "Currently, the traditional approach is to acquire decision making skill through experience. Only recently have efforts such as Cockpit Resource Management training and work by researchers such as Dr. Jensen shown that decision making is a skill that can be taught."

3   We may have improved the situation. According to the 1992 Transport Canada SAFE report, Personnel accounted for 69% of the total deficiency factors. Of those, 31% involved pilot judgement. The 1992 report is cumulative and covers the analysis of about 4000 occurrences since 1984.

4   Captain Heino Caesar, who recently retired as Lufthansa's Safety Pilot and General Manager, Flight Operations maintains an internationally recognized data base of hull losses. In the British Airways magazine, Flight Deck (Winter Edition, 1993) he reports that the possibility of flight crews being able to avoid losses has reduced to 54% in 1991 as compared to the long term statistic of 76%. However, Captain Caesar tempers his enthusiasm for our progress with a warning that we have developed new problems. "It is a matter of concern to me that the working environment for cockpit crews has become more hostile."

5   Captain Caesar concludes that, during 1991, 83% of the "human" failures, i.e. those of cockpit crews, were classified as H3: failures in judgement and/or decision making, and inappropriate aircraft handling.

6   United States participation was funded by the Buehler Aviation Research, Inc., directed by the Aircraft Owners and Pilots Association Air Safety Foundation with assistance from Events Analysis, Inc., and the Department of Aviation, Ohio State University. Canadian participation was funded by Transport Canada. Support was provided by the Royal Canadian Air Cadets who participated in the validation studies.

7   Principals were, Russel S. Lawton for AOPA, Mr. Ludwig Benner for Events Analysis and Dr. Richard S. Jensen of Ohio State University.

Canadian participation was headed by Dr. Georgette Buch, Transport Canada. Mr. Al Diehl, then of the FAA, provided considerable input and support to the project.

8 See p. 14 of the **Commercial Pilot Survey (1991) Levels III to VI Air Carrier Operations,** conducted by the Accident Prevention Branch of the Transportation Safety Board of Canada.

9 As Dr. Buch pointed out in 1984, "The ultimate measure of their (judgement concepts) effectiveness must be the reduction in the current high proportion of judgement-related accidents. "

10 Canadian lesson plans were developed by Al Diehl, then of the FAA and Dr. Georgette Buch of Transport Canada.

5 Cockpit crises and decision making: implications for pilot training

Maureen A. Pettitt, PhD, Western Michigan University

Crisis and group decision making

The aviation community has long realized that the effective performance of cockpit crews is essential to aviation system safety. Early research in the area of flight crew performance focused primarily on skills acquisition and retention, perceptual requirements, and physical stress. Much less attention was given to the psychosocial aspects of the cockpit environment. Air transport accident analyses and related research during the past decade have, however, produced convincing evidence that pilot training and evaluation systems must address the crucial dimension of crew interaction and decision making in the cockpit (Foushee 1984; Ruffell Smith 1979; Sears 1986; Wheale 1984).

In recent years many airlines have initiated training programs to encourage effective cockpit resources management (CRM). Although CRM programs vary, they are essentially designed to educate pilots (and, more recently, cabin and ground crews) about the importance of interpersonal relations, communication skills, synergistic activity, and participatory decision making to safe flight operations--training quite in contrast to the traditional focus on the technical means of accomplishing the goals of flight operations rather than the process.

Subsequent, evaluative research has supported the notion that CRM training can improve cockpit performance and, further, suggests that performance-related attitudes are significant predictors of crew coordination in line operations (Helmreich, Foushee, Benson, and Russini 1986).

However, research also indicates that crew members' lack awareness of the deleterious effects of stress and have unrealistic attitudes about their personal vulnerability to stress (Helmreich 1984). Some researchers caution that during a crisis situation the crew is likely to revert to prior well-learned behaviors rather than the concepts espoused by CRM (Hackman 1987b).

Alternatively, decision making is likely to be more successful in cockpit

31

crises if (1) group norms encourage shared information and responsibility for decision making (Foushee 1984; Foushee and Helmreich 1988); (2) the crew has confidence that a satisfactory solution exists and believes that there is sufficient time available to search for and evaluate alternative courses of action (Janis and Mann 1977); (3) the crew evaluates and utilizes available resources to develop alternatives, strategies or new resources, rather than relying upon existing strategies and resources to resolve the situation (Hackman 1987a); and (4) individuals are trained for cognitive flexibility under adverse conditions (Dutton 1981).

It is expected that over a period of time CRM training--if an accepted and customary component of initial and recurrent training programs--will encourage behaviors which lead to more effective coordination and decision making in critical or crisis situations (Hackman 1987b; Helmreich 1984; Helmreich and Wilhelm 1987).

Description of the research

The purpose of the study was to examine pilots' perceptions about three constructs central to decision making in cockpit crisis situations--the perception of crisis, sense of urgency, and response rigidity. The survey materials included a crisis scenario, a two-part, nineteen-item questionnaire and a background information sheet. The scenario and questionnaire items had been pre-tested during personal interviews conducted with twenty-four Denver-based line pilots employed by a major airline.

The subjects of the study were Los Angeles-based line pilots from three major airlines. Six hundred and fifty-seven surveys were distributed. One hundred and eighty-five usable surveys were returned and used in the analysis. The subjects represented a relatively broad cross-section of the pilot population (i.e., 46% were captains, 33.5% were first officers, and 20% were second officers; 38% were not yet forty years old, the other 62 were; 67% had no formal CRM training while the other 33% had).

The concept of crisis was measured in two ways. The perception of crisis was determined by asking pilots if they believed that this crew was in crisis situation (Question 1, Part I). They were asked to respond on a Likert-type scale numbered 1 (strongly agree) through 7 (strongly disagree). In Part II of the questionnaire pilots were asked to rate five crisis characteristics of the scenario on a Likert-type scale numbered 1 (low) to 9 (high). The second measure of crisis is a combination of the mean ratings of four crisis characteristics--level of threat to the safety of the flight, level of situational uncertainty, availability of decision-relevant information, and the level of tension.

The perception of urgency--the perceived time available in which to search for and evaluate alternative courses of action--was measured by combining the responses to three questions in Part I in addition to the rating of the fifth crisis characteristic, level of time pressure, in Part II of the questionnaire.

Response rigidity is characterized by the restriction of participation, limiting the search for and evaluation of viable alternatives, and adherence to/reliance on authority and procedures. Nine questions in Part I of the questionnaire were used to determine response rigidity.

It was hypothesized that (1) the perception of crisis would have a positive correlation to the ratings of the crisis characteristics, (2) the higher the rating of the situation as a crisis, the higher the response rigidity, and (3) a high sense of urgency would result in high rigidity.

Results of the study

Ninety percent of the pilots surveyed indicated, at some level of agreement, that the crew in the scenario was in a crisis situation. This crisis perception positively correlated with their ratings of the level of threat, uncertainty, information availability and tension--characteristics commonly attributed to crisis situations.

It was anticipated that the effects of the pilot personality, hierarchical task/role structures in the cockpit, and pilot training procedures--which have, traditionally, emphasized individual, mechanistic responses to emergencies-- would lead to a high response rigidity when the perception of crisis was high. Response rigidity is characterized by reluctance to engage in participatory decision making or to search for and evaluate alternative courses of action, and reliance on the captain's capabilities, deference to his or her authority, and reliance on standard operating procedures.

The hypothesis that a higher perception of crisis would result in higher response rigidity was not supported by the data. The results indicated that a higher perception of crisis resulted in a lower rigidity score (r=-.18). Further, pilots had, overall, a lower rigidity score than expected.

It was also expected that a higher sense of urgency would result in higher response rigidity. Previous studies of decision making suggest that a high sense of urgency (the belief that there is little time to search for and evaluate alternative courses of action) may evoke dysfunctional decision making behavior characterized by the restriction of information, authority, and participation. The hypothesis that high urgency would result in high response rigidity was supported by the data. The higher the sense of urgency the higher the response rigidity (r=.25).

The findings indicate that the perception of crisis and response rigidity are negatively related (high crisis/low rigidity, low crisis/high rigidity) and the sense of urgency and response rigidity are positively related (low urgency/low rigidity, high urgency/high rigidity). Rigidity scores were not significantly different as a result of a high or low perception of crisis, however, pilots with a low sense of urgency had significantly lower rigidity scores than pilots with a high sense of urgency. These results suggest the interpretation that sense of urgency--the time component of crisis--drives decision making behavior, more so than the perception of crisis.

While the results of this study were not conclusive, they strongly suggest the possibility that the high crisis perception/low urgency/low rigidity pattern may be an optimal approach to crisis decision making. First, the decision maker recognizes the situation as a crisis. As a result of this awareness the decision maker experiences mild but "helpful" stress and is, consequently, motivated to act on the situation. The low sense of urgency--the belief that there is sufficient time to search for and evaluate alternative courses of action--encourages flexibility with respect to roles, responsibility, participation, and procedures.

Conversely, pilots exhibiting the high crisis/high urgency/high rigidity pattern may be behaviorally similar to people who suffer from the high, debilitating stress which ultimately inhibits performance. They may resemble the hypervigilant decision maker described by Janis and Mann (1977) whose high arousal state results in impaired cognitive functioning and narrowed time perspectives. The study similarly suggests (although less conclusively) that a high, or at least moderate, perception of crisis is antecedent to the low urgency/low rigidity response pattern. The present research makes a stronger case for low sense of urgency as antecedent to flexible, participatory decision making.

Implications for pilot training

The research lends additional support to the findings that cockpit resources management training improves attitudes toward crew coordination and decision making (e.g., Helmreich 1989). Pilots who had attended formal CRM training programs exhibited a significantly lower sense of urgency and significantly lower response rigidity than pilots who had no formal CRM training.

The results of this study also provide some possibilities for expanding or modifying content in pilot training programs in industry and those used in colleges and universities. While there are implications for several concepts central to most CRM training programs--decision making, communication, stress management (U.S. Federal Aviation Administration 1989)--the concept of situational awareness is most relevant to the present discussion.

It has been argued that an accurate assessment of crisis is a necessary first step toward crisis resolution (Billings, Milburn and Schaalman 1980). In aviation, this notion is embodied in the concept of situational awareness.

34

Situational awareness is generally considered to be the accurate perception of the factors and conditions that affect the aircraft. In other words, a pilot who is situationally aware has made an accurate assessment of reality. However, the concept of situational awareness as currently conceived focuses primarily on information processing and communication (e.g., Nagel 1988) with little reference to temporal structure or awareness.

The results of this study suggest that it would be beneficial to expand the concept of situational awareness to include (a) the concept of crisis and (b) the accurate assessment of decision time available in critical situations (as opposed to relying on an internally-perceived time frame). Pilots should be aware that in a crisis time perspectives are constricted and short-term goals and consequences tend to be overemphasized.

An expanded concept of situational awareness is applicable to simulator training as well as ground training. Debriefing sessions could include evaluation of how the crew assessed the decision time available, the accuracy of that assessment, and how time perspectives affected the processes used to resolve the crisis. This approach is not limited to the full-motion, high-fidelity simulators utilized by major airlines. Such training might be accomplished just as effectively in low-fidelity simulators or using interactive video workstations and video recordings of behavior (e.g., Foushee and Helmreich 1988).

References

Billings, R. S., Milburn, T. W., & Schaalman, M. L. (1980), 'Crisis perception: A theoretical and empirical analysis', *Administrative Science Quarterly, 25*, 300-15.

Dutton, J. E. (1986), 'The processing of crisis and non-crisis strategic issues', *Journal of Management Studies, 23*, 501-517.

Foushee, H. C. (1984), 'Dyads and triads at 35,000 feet', *American Psychologist, 39*, 885-893.

Foushee, H. C., & Helmreich, R. L. (1988), 'Group interaction and flight crew performance', In E. L. Wiener & D. C. Nagel (eds), *Human factors in aviation*, 189-227. San Diego, CA: Academic Press.

Hackman, J. R. (1987a), 'The design of effective work teams', In J. W. Lorsch (ed), *Handbook of organizational behavior*, 315-342. Englewood, NJ: Prentice-Hall.

_____. (1987b), 'Group level issues in the design and training of cockpit crews', In H. W. Orlady & H. C. Foushee (eds), *Cockpit resource management training: Proceedings of the NASA/MAC Workshop* (NASA CP-2455), 23-39. Moffett Field, CA: NASA-Ames Research Center.

Helmreich, R. L. (1984), 'Cockpit management attitudes', *Human Factors, 26*, 583-589.

_____. (1989), 'Evaluating the effectiveness of cockpit resource management training', Washington, D. C.: 1989 Human Error Avoidance Techniques Conference. Photocopied.

Helmreich, R. L., Foushee, H. C., Benson, R., & Russini, R. (1986), 'Cockpit management attitudes: Exploring the attitude-performance linkage', *Aviation, Space and Environmental Medicine, 57*, 1198-2000.

Helmreich, R. L., & Wilhelm, J. A. (1987), 'Evaluating cockpit resource management training', In R. S. Jensen (Ed.), *Proceedings of the Fourth International Symposium on Aviation Psychology* (pp. 440-446). Columbus, OH: Ohio State University Department of Aviation.

Janis, I. L., & Mann, L. (1977), *Decision making: A psychological analysis of conflict, choice, and commitment.* New York: The Free Press.

Miller, K. (1963), 'The concept of crisis', *Human Organization, 22*, 195-201.

Nagel, D. C. (1988), 'Human error in aviation operations', In E. L. Wiener & D. C. Nagel (Eds.), *Human factors in aviation*, 263-303. San Diego, CA: Academic Press.

Ruffell Smith, H. P. (January, 1979), *A simulator study of the interaction of pilot workload with errors, vigilance and decisions* (NASA Report TM-78482). Moffett Field, CA:NASA-Ames Research Center.

Sears, R. L. (1986)' *A new look at accident contribution and the implications of operational and training procedures.* Seattle, WA: Boeing Commercial Aircraft Company.

U. S. Federal Aviation Administration, (December, 1989), *Cockpit Resources Management Training*, Advisory Circular, 120-51. Washington, D. C.: Federal Aviation Administration.

Wheale, J. L. (1984), 'An analysis of crew coordination problems in commercial transport aircraft', *International Journal of Aviation Safety, 2*, 83-89.

Zajonc, R. B. (1965), 'Social facilitation', Science, 149, 269-274.

6 US air carrier emergency evacuation events: necessary or negligent aeronautical decisions?

Prof. Michael K. Hynes, ATP, CFI, IA, A&P, Western Oklahoma State College, Altus, Oklahoma

Introduction

Background

On the average, US Federal Aviation Administration (FAA) Part 121 Air Carrier crashes occur in the USA less than twice per year. It is not uncommon for emergency passenger evacuations to take place as a result of these crashes. Because of the wide variety of circumstances surrounding these accidents, and how emergency passenger evacuations are carried out, it is difficult to obtain statistically significant information on these events to analyze the many human factor aspects of emergency evacuations.

This study shows that every 14 days, an emergency evacuation event (EEE), which is not "crash" related, does take place on a large Air Carrier aircraft. By analyzing these non-crash related evacuations, information for improving passenger safety during evacuation events can be obtained.

The reasons for non-crash related emergency evacuation events range from bomb threats, to observations of "smoke' from air conditioning ducts, to flat tires during taxi and various other causes. In some cases, evacuations are improperly initiated by over zealous passengers.

Purpose

The purpose of this study was to document the need for improvements in passenger safety when circumstances arise which *suggest* the need to consider an emergency evacuation. A research objective was to acquire information on Emergency Evacuation Events (EEE) that took place which were *not* associated with crashes. This information was used to answer the questions: (1) Can the number of these events be reduced? (2) Can the number of passenger injuries that result from these events be reduced? (3) Should all EEE be reported to an agency, such as the National Transportation Safety Board (NTSB) so safety information and trend data can be analyzed? (4) Can management policies and/or training programs, of airline and airport operators, be modified to reduce non-crash related EEE and reduce the number of passenger injuries sustained during evacuations?

The Study

In the initial phase of this study, the FAA and the NTSB were contacted to discuss the concept of the research. Because the computer data bases of both of these organizations indicated that emergency evacuation events were infrequent and limited data existed on passenger difficulties when the events did take place, the interest of the FAA and the NTSB in this research was low.

Procedures

Using the non-occurrence of evacuation events as a hypothesis, it was determined that there was a need to check the accuracy of the FAA and NTSB data bases on EEE to confirm or disprove the hypothesis, before further research on the subject would be undertaken.

Searches were made of the FAA Oklahoma City Aeronautical Center and the NTSB Washington, DC data bases. An earlier NTSB study related to this subject, plus data from the National Aeronautics and Space Administration Aviation Safety Reporting System was also used. The total number of reports exceeded 250,000. The information from these four sources was further supplemented by a review of a limited number (n=20) of newspapers for articles on this topic along with a small sample of litigation activities of passengers who claimed injuries resulting from EEE (n=9).

Written and phone contact was made with the management of 49 airports, including the 25 busiest. These airports accounted for approximately 85% of air carrier operations and passenger enplanements. A standard research instrument (questionnaire) was used to obtain data on all evacuation events.

A final step of the study was to interview FAA and NTSB employees who had access to, or interest in the topic of EEE. These discussions were supplemented by interviews with airline management, ground crew personnel and flight crews, both pilots and cabin staff (n=100).

To analyze non-crash related EEE from a management perspective, the contents of several different Air Carrier Operations Manuals (n=7) were reviewed for information on *non-crash related* Emergency Evacuation Events.

Limitations of the study

The limits of time and resources reduced the ability to further analyze the newspaper and litigation information data bases. While a number of potential emergency evacuation events were found in both of these resources, an analysis of these events in sufficient depth to include them in this study was not possible. Therefore, not all non-crash related EEE were identified and studied.

The time period used for this study was from 01/01/88 to 12/31/92. Data from earlier periods was available, but not from multiple sources since requesting verification of older data created a burden for airport operators. The impact of "wide-body" aircraft on passenger safety during EEE was important to the study outcome, therefore a recent five year period was selected as a representation of the air travel conditions being studied. Data after 12/31/92 was available, but delays in entering information into FAA and NTSB data bases, limited its availability and precluded its use.

Assumptions

It was assumed that when any emergency evacuation event was reported, the event did in fact take place. It was also assumed that the five year period selected was a representative sample of US domestic Air Carrier activity which would reveal the required information for this study.

Results of the Study

In direct contrast to the NTSB and FAA statements regarding the infrequency of emergency evacuations, during the five year period studied, on an average of every 14 days an evacuation event did occur on a "large" aircraft that was being operated by an Air Carrier. In the 60 month period, at least 145 evacuations took place, only 5 of which were "crash" related. On ten other occasions, damage to the aircraft did occur due to landing gear malfunctions. An almost equal number of evacuations took place on small and medium size air carrier aircraft. Data from these events was not included in this study.

The NTSB data on emergency evacuation events was incomplete and/or had an accuracy probability of less than 10%. One NTSB data base showed 3 events, another only 17, while the FAA data base had reports on 88 of the events. The FAA data had a range of 42% to 75% accuracy, depending on the year reported and selected for analysis.

The response rate, to the airport operators research instrument was 95.9%.

While the passenger count during these events was incomplete, 14,224 persons were identified as being evacuated during 85% of the events. A total of 324 injuries were reported during non-crash events, however this information was missing in 12.8% of the evacuations reported. Using these statistics as a base, during non-crash evacuations, a potential of over 16,000 passengers were evacuated and 400 injured during the five year period studied.

The number of passenger initiated, non-crash related EEE, was 6.9%. The most common reason for a passenger initiated EEE was a perception of a fire external to the aircraft associated with the start of the auxiliary power unit (APU) or a "hot start" of an engine that was in view of a passenger. Condensation "smoke" and unusual odors in the aircraft air-conditioning system also caused unwarranted passenger concern. The perceived fire threat was almost always found to be a "false alarm."

The other significant finding of the study was the automatic, rote or reflexive actions of the cockpit crew to non-crash related EEE decisions. It frequently appeared that little aeronautical decision making or any cognitive process was undertaken by the flight crew prior to "ringing the bell" to evacuate the aircraft in situations where no crash had taken place. In these cases, no opportunity was available for the cabin crew to prepare themselves or the passengers for a very traumatic event, the emergency evacuation of a large aircraft. Under these circumstances, passenger injuries were always reported.

Whenever a wide body aircraft was evacuated, there was a 100% probability that passenger injuries would be reported.

In many cases, airport personnel were not informed in advance that an EEE was going to be conducted.

In some cases, evacuations had no support by airport or airline ground personnel. Often, after the evacuation, passengers were told to remain close to the aircraft. This canceled the safety benefits gained by the evacuation, and in some cases, actually increased the potential for harm to the passengers.

When Crash Fire Rescue (CFR) personnel were present during an EEE, Communication between the aircraft and CFR staff was usually non-existent.

Conclusions of the Study

Emergency evacuation events occur more frequently in Air Carrier operations than the FAA, NTSB or others seem to realize. For every "crash related" event, 29 non-crash related events occur just on "large" aircraft. When the total number of passengers subjected to these events is taken into consideration, over 3,900 persons in some years, this situation presents a safety hazard that is worthy of further study and correction.

The reason for the failure of the FAA and/or the NTSB to be aware of the high frequency of emergency evacuations is based on two facts. First of all, the regulations apparently do not call for the reporting of these events, and secondly, the adverse publicity of these events, from the viewpoint of airport operators, manufacturers, and airlines, discourages reporting evacuations.

Even when evacuation slides fail, there is some debate whether or not this information has to be reported to the FAA under FAR 121.703, *Mechanical Reliability Reports*. While section (a) (17) addresses "Emergency evacuation systems" the regulation refers to only "during flight" failures. Part 121.703 (b) states *during flight* "means the period from the moment the aircraft leaves the surface of the earth on take off until it touches down on landing." This automatically excludes almost all of the non-crash related EEE. Section 121.703 (c), which might be a "catch all" reporting requirement of failures, requires reports only on items which "may endanger the safe operation of an *aircraft*." This seems to have been interpreted to exclude events which might endanger passengers but *not* the aircraft.

The breakdown in communications between the cockpit and cabin crew was a noted situation in many events. The usual problem was evacuation orders being given to cabin crews without any pre warning that the event would take place. Written reports and sworn testimony confirm the confusion that takes place on large aircraft when emergency evacuations are carried out without passenger briefings being conducted prior to the event. In a few cases, cabin crews initiated the evacuation without the knowledge of the flight crew and the aircraft was "still in motion" as the cabin crew began the evacuation. This was a common problem if passengers improperly initiated the evacuation process.

Because of the serious nature of passenger initiated evacuations, the standard passenger announcement and briefing cards should warn passengers, that *unless an actual accident takes place*, they are not to respond to evacuation instructions from persons other than the flight or cabin crew. If passengers responded *only* to crew evacuation instructions, it would greatly diminish the frequency of "out of control" events improperly initiated by passengers.

Before the "jet age", when large radial engines of commercial airliners were started, smoke and fire would often be seen, especially at night. The standard cabin announcement would warn of this event and request that passengers not react adversely to this sight. It would be cost effective to modify today's cabin announcements to include a similar warning that signs of smoke and fire may be present during APU or engine starts. Odors and condensation "smoke" from air conditioning systems are also reasons for premature evacuation actions by passengers. Information should be added to passenger instruction cards in an effort to reduce the likelihood of passenger concern when these potential signs of fire take place..

At very little cost, changes can be made in policies and training programs of both airline and airport operators to reduce the frequency of EEE, and also reduce the potential for injuries during evacuations *which are found to be necessary*. Better aeronautical decision making in non-crash related EEE, and in crew briefings before flight, would significantly add to passenger safety.

In view of the fact that 100% of the EEE, that take place on wide-body aircraft result in reported passenger injuries, the present method of testing aircraft evacuations under FAR 25.803 must not be providing the level of safety the public expects and is entitled to receive. During recent certification of the MD11, "volunteer passengers" were injured during evacuation testing. The FAA allowed a relaxing of the test conditions versus considering the possibility of design problems with the aircraft exit systems. This was not a prudent solution to the passenger injury problem during these test evacuations.

Communication between air crews and CFR personnel, prior to and during EEE, would increase passenger safety. The international common emergency frequency of 121.5 should be used for these events.

Airport management should select preferred aircraft parking sites prior to evacuations. These sites would depend upon the circumstances of the event, such as bomb threats. This will result in a higher level of passenger safety.

Because of the perceived lack of need, the human factors aspect of EEE has not been researched in depth. Either acting on its own, or through the use of contractors, the FAA, NTSB, or others in the aviation community, need to obtain more information on EEE, especially events that are non-crash related.

The results of this study has "public benefit" by calling attention to the need to enhance the safety of air travel by decreasing the frequency of non-crash related EEE. It also suggests improvements to the management policies and training programs of airline and airport personnel, as applied to both crash and non-crash related EEE. Adopting some or all of these suggestions should reduce passenger injuries when passenger evacuations are conducted.

Recommendations

Based on the findings of this research, the following recommendations should be considered for adoption by the appropriate parties:
a. There is a need for detailed information on *all* emergency evacuation
     events. Mandatory reporting should be required by FAR, Part 121.
b. Passenger announcements and briefing cards need to be modified to
     include information on the signs of a false fire and for passengers to
     respond only to flight and cabin crew instructions to evacuate.
c. Review aircraft certification standards, FAR Part 25.803.
     i. Wide body aircraft have 100% injury probability. Certification testing,
        as now conducted, does not seem to be realistic.

41

ii. There is a need for a second "able bodied assistant" assigned as a backup at each exit position in the cabin and on the ground.
d. Communication needs to be improved and standardized:
    i. In advance of the event: with all members of the crew (cockpit and cabin); with airport personnel, CFR, Air Traffic Control; airline management, and with the passengers.
    ii. During the evacuation: by utilization of common radio frequencies with CFR personnel and the aircraft (on 121.5?).
e. Airline Policies on non-crash related EEE need to be modified.
    i. When are non-crash related EEE to be conducted?
    ii. What are the notification procedures, to whom and when?
    iii. Better Crew Resource Management (CRM) Training in Aeronautical Decision Making and joint EEE training (cockpit and cabin crews).
    iv. Better pre-flight briefings on non-emergency but abnormal situations.
f. Airport policies on non-crash related EEE need to be updated, and
    i. made available to airline personnel, both management and operations;
    ii. notification requirements need to be established;
    iii. preferred locations where EEE should take place need to be identified;
    iv. improve CFR response to EEE (stairs and other equipment).
g. Additional studies are needed on the human factors aspects of emergency evacuations to reduce the number of events, to reduce the number of passenger injuries during the events, and to reduce the number of events initiated by passengers as a reaction to false perceptions of fires.

References

Accident/Incident Data- Emergency Evacuation-FAR Part 121 Operations, (24MAR93), Report No. P3-03-0206, FAA, Oklahoma City, OK.
Air Carrier Cabin Evacuation Safety Recommendations. (09/23/68-06/24/92) NTSB, Washington, DC.
Air Carrier Emergency Evacuation Reports. (Search Request No. 3223) (1993). NASA, Aviation Safety Reporting System, Mountain View, CA.
Air Line Pilots Association. (1989), *The ALPA Guide to Accident Survival Factors.* Engineering and Air Safety Department, Herndon, VA.
AIRCRAFT EMERGENCY EVACUATION: A Method for Evaluating Devices, Procedures, and Exit Provisions. (1951) Civil Aeronautics Administration, Office of Aviation Safety. Washington, DC.
Chandler, R., George, M., and Pollard, D. (1987), *Usability of Transport Aircraft Emergency Exits.* FAA Civil Aero Medical Institute, Oklahoma City.
Evacuation Incident Reports, (09/23/68-05/01/93) NTSB, Washington, DC.
Handbook Bulletin 91-36. (Manual 8400.10, Change 6, Appendix 3). FAA, Washington, DC.
Muir, H., Marrison, C., and Evans, A. (1989), *Aircraft evacuations: the effect of passenger motivation and cabin configuration adjacent to the exit.* CAA Paper 89019. Civil Aviation Authority, London.
Ostrich, R., Starr, W. (1956), *Passenger Escape from Aircraft Employing the Inflatable Slide.* Military Air Transport Service, Andrews AFB, Washington.
SPECIAL STUDY: Safety Aspects of Emergency Evacuations From Air Carrier Aircraft. (1974)Report Number AAS-74-3. NTSB, Washington.

7 Operator decision making: information on demand

Marion Kibbe, Naval Air Warfare Center Weapons Division, China Lake, California
Edward D. McDowell, Oregon State University, Corvallis, Oregon

Introduction and background

The requirement to manage and interpret new sources of information available in attack aircraft threatens to increase pilot workload to intolerable levels. For this reason, automatic target recognition systems (ATRs) are under development which will assist pilots to detect, track and identify targets. The pilot will accept or reject ATR recommendations based upon the known or estimated reliability of the ATR and its congruence with the additional information available in the cockpit.

A series of laboratory experiments (Kibbe, 1992; Weisgerber, 1990) has examined operator target identification performance using both simulated and FLIR imagery of ships aided by simulated ATRs. The ATR output was one or more possible ship identifications plus a figure of merit (FOM) which expressed the ATR's probable accuracy.

Results showed that system performance (the operator using an ATR) might be better, equal, or worse compared to the performance of the autonomous ATR or the unaided operator. Operator target identification aided by an ATR was almost always superior to unaided operator performance; gains exceeded 15% in some cases. Synergistic operator/ATR performance was also found in several instances; that is, the operator/ATR together performed significantly better than both the unaided operator and the autonomous ATR. In some situations, operator/ATR performance equaled the performance of whichever

43

was the better of the two individual components. Sometimes the inclusion of a human operator harmed system performance: the operator/ATR together performed less well than the better component which was the ATR.

The tentative conclusions from these experiments were: 1) synergy occurred when the performance levels of the unaided operator and the autonomous ATR were very similar; and 2) the human operator harmed system performance when his unaided performance was substantially below the performance of the autonomous ATR. One of the goals of this experiment is to further examine operator/ATR system performance as a function of both the absolute and the relative performance differences between the autonomous ATR and the unaided operator. An understanding of the dynamics of this relationship should lead to design guidelines for operator/ATR systems and for the interface between them.

Method and Design

Simulated incoming imagery of 15 ships was developed by taking broadside ship pictures, and representing each profile by 60 vertical bars. Good quality imagery preserved the correspondence between the height of the bars and the height shown in the pictures, while poor quality imagery did not. The same simulated imagery was provided both to the ATR and the experimental subjects, but the quality was varied independently so that both could be shown good imagery (the SgAg image quality condition), both could be shown bad imagery (SbAb), or one could be shown good and the other poor imagery (SgAb or SbAg). To make its ship identifications, the ATR correlated the imagery shown to it with the good quality ship images stored in its library. The output of the ATR was either one or five possible ship identifications, and the magnitude of the correlation between the incoming imagery and the library images served as the basis for providing FOMs for each recommendation.

Imagery of any one of 15 ships was shown to subjects in five different experimental conditions: 1) The Unaided condition in which operators saw the ship and received no ATR information; 2 & 3) The Given-1 and Given-5 conditions in which along with the ship image, an ATR provided either 1 or 5 possible ship identifications each with a FOM; and 4 & 5) The On Demand-1 and On Demand-5 conditions which are like the two Given conditions, except that subjects had to request identification(s) by pressing a function key and had to request FOM(s) by pressing a second function key.

The five experimental conditions were given in counterbalanced blocks and within each block the image quality varied. There were 15 subjects who, after training in ship recognition, participated in the experiment (see Table 1).

Table 1 Experimental design

Image quality	Unaided	Given-1	Given-5	On Demand-1	On Demand-5
SgAg	x	x	x	x	x
SgAb		x	x	x	x
SbAg		x	x	x	x
SbAb	x	x	x	x	x

Results

Unaided vs. Aided Ship Recognition.   Unaided operator performance identifying the ships averaged 83% correct while in the aided conditions (the four Given and On Demand conditions) it averaged 92% correct (p< 0.001). Analyses of response times (RTs) showed that aiding did not alter RT: unaided RT was 11.4 seconds and aided RT was 10.9 seconds.

Information Given vs. Information On Demand.   There were no differences between the four aided conditions. Operators were equally correct whether the ATRs gave 1 or 5 recommendations or whether ATR output was given automatically or on demand. Accuracy varied only as a function of image quality (p < 0.001).   The analyses of RT gave similar results: there were no speed differences between the 4 aided conditions, however RT varied significantly with image quality (p < 0.001). Responses were approximately 1 second faster with good imagery (Sg) than with bad (Sb).

Analyses of the On Demand conditions only.   Operators asked for ATR output, either a recommendation, or a FOM, or both, on 64% of the trials. When they asked for ATR output, 81% of the time they requested both an identification and a FOM. Subjects sought aid more frequently when they saw bad imagery than when they saw good (p < 0.001) and when the ATR gave 5 recommendations rather than only one (p < 0.07).   Subjects did not perform better when aided by the ATR, in fact they performed slightly, but not significantly worse.

Finally, a scan of the records of the 14 subjects showed that people used different strategies in requesting ATR information: two subjects requested ATR output on every trial; while one subject never requested aid. The other subjects ranged between these two extremes. Spearman correlation of ranks between the 1-and 5-recommendation conditions, based upon the frequency of aid requests, showed that those who frequently asked for aid in one condition also sought aid in the other (r_s = +0.87, p< 0.001). Correlations of these ranks with accuracy of ship identification rankings were not significant, indicating that different strategies of ATR use did not lead to better performance.

Synergy. Figure 1 shows a comparison of the unaided operator, autonomous ATR, and operator/ATR system performance for the single recommendation condition. Synergy was defined as system performance that is significantly greater than the performance of better of the two individual components. In the SgAg condition, the performance of the operator/ATR was not significantly better than that of the unaided operator, and so we regard this as a pattern where system performance was equal to the performance of the better component. The same was true in the SgAb and SbAg image quality levels. In the SbAb condition the system performance was synergistic; the operator/ATR was significantly better than the unaided operator (p < 0.004).

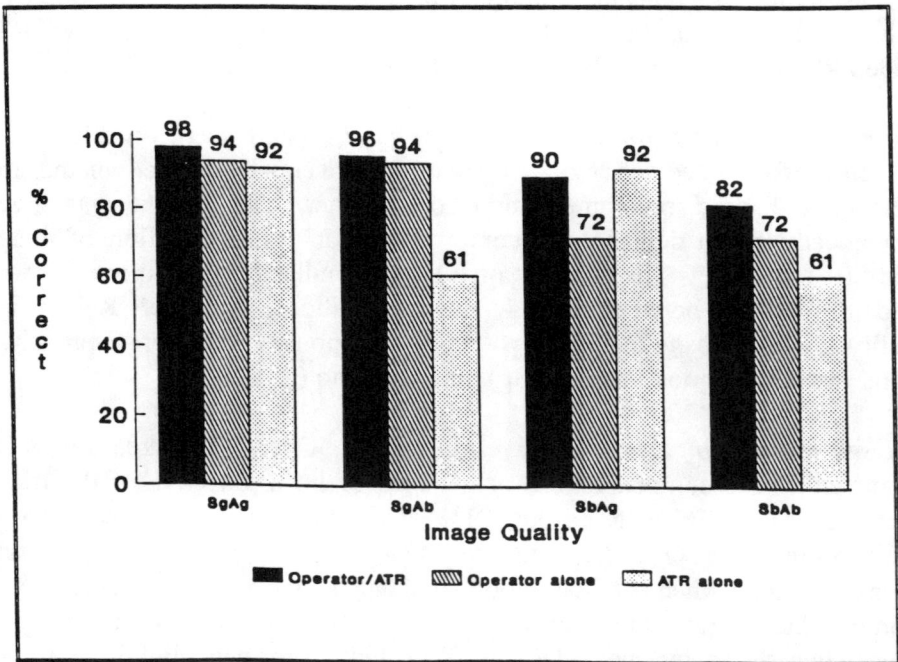

Figure 1 System vs. component performance, one ATR recommendation

The 5-recommendation condition is shown in Figure 2. In this condition, both the SgAg and SbAb conditions were synergistic: the operator/ATR performed significantly better than the unaided operator when all had good imagery and also when all had poor imagery. In the SgAb and SbAg conditions, system performance was equal to the performance of the component that was given the good imagery.

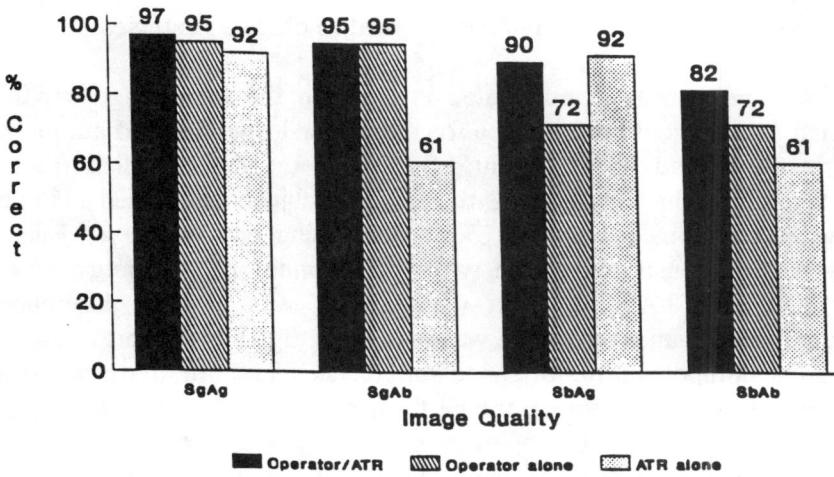

Figure 2 System vs. component performance, five ATR recommendations

Discussion and conclusions

The results showing that aided performance is superior to unaided and that both accuracy and RT are better with good image quality are completely consistent with previous findings. The lack of differences in accuracy between the Given and On Demand conditions suggests that when the subjects did not want or need the ATR output, they ignored it and it did not alter their performance. In the On Demand condition, the fact that subjects more frequently asked for help when they could get 5 recommendations suggests that subjects may use ATR output to confirm a ship identification hypothesis that they generated from looking at the imagery. They preferred 5 ATR recommendations to 1 recommendation since they would be more likely to have their hypothesis reinforced.

Aiding neither added to nor decreased response time; this was true both for the Given and On Demand conditions. Since aid was requested more often than not in the On Demand condition, the lack of an RT difference must reflect both that the request process was very easy and fast, and that it was effectively shared with the time taken to observe the imagery and form a hypothesis about its identity. It was surprising really that more people did not ask for aid on every trial. This lends credence to the notion that people vary in their styles of

47

interaction with machines and automated systems. While this experiment did not show one strategy superior to another, we need to know more about decision strategies, if they change over time, if they can be taught, and whether performance is altered if an individual is forced to change strategies.

Synergy. Synergy was demonstrated in three of the four image conditions which had nearly equal performance for the unaided operator and autonomous ATR (the SgAg and SbAb conditions). While the design called for equal performance from the components, the unaided subjects performed 11% better than the autonomous ATR when both were using bad imagery. Even so, synergy was demonstrated in the two SbAb conditions. Synergy was not observed in the SbAg condition where there was a 20% difference in component performance, and none was seen in the SgAb condition where there was a 34% component performance difference. In both of these, system performance was nearly equal to the performance of the better component, the ATR in SbAg, and the unaided operator in SgAb. Thus we can infer that while the performance of the components needs to be relatively equal to obtain synergy, there is tolerance of an 11% difference, but not a 20% difference. It is possible that there would not have been synergy if the ATR had been superior to the operator. If operators have a tendency to overvalue their own identifications relative to the ATR's, they will reject accurate ATR output more frequently than one would expect on a chance basis. When the ATR's performance is better than the human's, overvaluation of the human component would harm system performance and preclude synergism. So while large component performance differences can lead to synergy when the operator is better than the ATR, there is no evidence from this experiment that system synergy will be found when the ATR is superior.

In terms of absolute performance levels, we demonstrated synergy when component performance was very high in the 90% correct range, and also when it was lower in the 61-72% range. It would be of great interest to know whether two very poor components could produce a synergistic system.

References

Kibbe, M. and McDowell, E. (1992), 'Man-Machine Interface with Simulated Automatic Target Recognition Systems', *AGARD Conference Proceedings,* No.521, 28:1-10.

Weisgerber, S.A.,and Savage, S.F. (1990), *Operator Ship Classification Using an Automated Target Recognition (ATR) System in Conjunction With Forward-Looking Infrared (FLIR) Imagery,* China Lake CA: Naval Weapons Center Technical Publication 7101.

Part 2
ATC: AUTOMATED SYSTEMS

8 The effects of reduced partyline information in a datalink environment

Susan Infield, Aileen Logan, Leysia Palen, Elfriede Hofer and David Smith, Boeing Commercial Aircraft Company
Kevin Corker and Sandy Lozito, NASA Ames Research Center
Antonio Possolo, Boeing Computer Services

Anecdotal evidence exists that airline crews use "partyline" information (PL), i.e., information overheard from conversations between ATC and other nearby aircraft. Such information is considered useful for clearance anticipation, nearby weather or turbulence conditions identification, and overall situational awareness. Researchers have attempted to systematically validate these reports (Midkiff and Hansman, 1992) and to categorize and prioritize PL by type and value. The introduction of DL for air/ground communication has raised concerns about its impact on PL availability and monitoring, and its effects on flight crew situational awareness. The concerns are two fold. First, the number of verbal radio transmissions and the volume of PL will decrease as aircraft become DL equipped and DL mediated operations spread across flight phases. Secondly, a DL environment may reduce the flight crew's tendency to monitor PL. This latter concern derives from the potentially reduced need for crews to monitor communications for their call sign in a DL environment.

Some pilot groups want DL to provide compensatory information to maintain situational awareness with reduced partyline communication. This may be a costly and technically difficult challenge. The types of PL currently used and their effects on crew behavior must be empirically identified before full operational introduction of DL communication.

Partyline monitoring is very much a "hit or miss" activity. To overhear and use PL presupposes several conditions: the information is on an aircraft monitored frequency, the information is heard by the flight crew, the flight crew is sufficiently situationally aware to understand the information, and the crew takes appropriate action. If relevant information can be determined, DL provided PL could actually improve rather than reduce situational awareness.

We introduced the partyline manipulations discussed in this paper as part of a larger experimental protocol designed to examine multiple implications of DL use in terminal and enroute environments. The reported results are limited to the effects of the partyline manipulation crossed with the major manipulation of DL implementation type.

Method

Subjects We recruited fifteen current and qualified line crews from two major U.S. domestic airlines. Crew members flew in their customary positions (captain or first officer). These volunteer crews received no financial compensation for their participation beyond hotel accommodations and food during their stay in Seattle.

Design The experiment used a between-groups design with two independent variables. All crews flew in a single condition but experienced identical routes with the same set of clearances, enroute traffic, weather, and mechanical failures. Communication mode was the principal manipulation. There were three conditions: (1) voice communication - used normal voice radio channels for all clearances and company communications; (2) FMS integrated DL - used a two-way DL implementation on the forward control display unit (CDU) for all clearances and company communications. This implementation interfaced with the FMS and permitted direct loading of ATC directives into the system; (3) retrofit, non-integrated DL - used a two-way DL implementation which used the center CDU for all clearances and company communications. This implementation did not interface with the FMS or permit direct loading of ATC directives into the system. A second manipulation (the focus of this paper) involved the manipulation of PL. Researchers coded PL as either relevant or irrelevant to the subject aircraft. Relevant messages included turbulence reports; weather information; air and ground traffic conflicts; and handling information. The voice communication condition had a full set of scripted PL. The two DL conditions had a reduced amount of overall PL. This simulated an early environment with approximately 25% DL equipped commercial aircraft. The ratio of relevant to irrelevant partyline was the same for all communication conditions. The dependent variables of interest were: (1) crew awareness of relevant PL, and (2) crew response to relevant PL.

Scenario The detailed flight scenario contained ground and airborne traffic conflicts, weather advisories, and mechanical faults. Flights originated at San Francisco International Airport bound for Washington National. High fidelity, out the window visual scenes encouraged environmental scanning. Subjects encountered other aircraft during initial taxi and in flight. Inflight events included routine traffic advisories and one emergency traffic evasion event. The detailed scripts included all clearances to the subject aircraft, and all PL conversations between ATC and traffic during the flight. PL categories were: Traffic conflict, Call Sign Similarity, Turbulence, Weather, Route Information/handling, Windshear/handling.

We translated voice clearance scripts into DL format for the two DL conditions. A highly experienced, retired controller provided ATC support. Experimental personnel (all qualified pilots) provided traffic conversation. The scenario contained a mechanical fault which resulted in a diversion to Denver after an average of 140 minutes.

Equipment We used used a 747-400 engineering simulator with full visual and systems capability. Researchers constructed two DL implementations. One system, integrated into the FMS, used the forward CDUs, (both First Officer and Captain side) as the message display and composition device. A second system shared general design principles and functionality with customer proposed retrofit DL implementations. This DL system used the center CDU as the interface device. Both systems allowed the crews to compose clearance requests and free text messages for downlink to ATC or company representatives. A more complete description of these implementations is found in Infield, Corker et. al. (1994). Low light video cameras behind the crew and in the front corners of the simulator cab recorded video and audio activities. Data streams from the simulator recorded aircraft movement, all crew inputs, and timing data. A remotely located ATC station included a flight movement monitoring screen, voice and DL broadcast communication station, and a "pilot" manned traffic communication station.

Procedures Researchers randomly assigned subject crews to experimental conditions with the reservation that equipment problems in either of the DL conditions might re-assign a crew to an alternate condition. Researchers assigned the first officers principal responsibility for communications activities. First officers received training in one of the DL implementations. Training averaged about 35 minutes. Captains received similar briefings, but were not required to reach the same level of competence as first officers, who were fitted with helmet-mounted eye tracking equipment. Data collection began with the crews' preflight briefing and ended after touchdown in Denver.

Results and discussion

Researchers examined video and audio taped data of flight crew experiment performance and transcribed descriptions of crew behavior that occurred directly before, during, and after each partyline event. An analyst used video data to rate each event/crew combination for awareness. Other analysts reviewed the transcribed record and verified the ratings. Analysts judged crews as aware of PL if they did at least one of three things in a way which clearly related to the PL information: communicated with ATC; communicated among themselves; or performed an action. Analysts judged crews as unaware of PL if they did at least one of the same three things in a way which clearly indicated lack of PL awareness. The sections that follow describe the partyline events in context and the results of our awareness analysis for each event.

Traffic conflict

A. ATC cleared another aircraft to cross the active runway in front of our subject aircraft holding at the end of the runway for take-off clearance. Weather conditions were hazy with visibility above minimums. The crews received take-off clearance within 10 seconds of this transmission. Only one voice condition crew indicated awareness. Three integrated DL condition crews responded to the event. None of the retrofit condition crews indicated awareness. All crews continued with the take-off.

B. Crews followed another aircraft through a series of clearances and into final approach after a go-around event caused by windshear. The leading aircraft informed ATC that they could not make the turn onto the cleared taxiway and would have to go all the way to the end of the runway. A side-step clearance from ATC to our subject aircraft followed this event. Only one voice crew anticipated the change in landing clearance after hearing the PL. Three crews from each of the DL conditions responded to the information.

Figure 1 shows the results of these analyses and the go-around event described under *Windshear*.

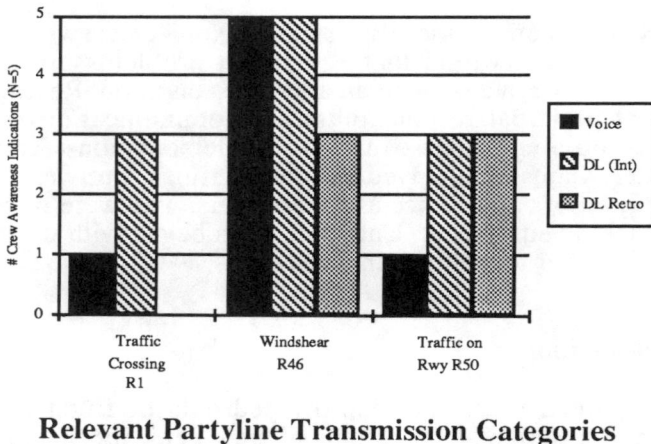

Relevant Partyline Transmission Categories

Figure 1 Terminal area safety transmissions

Call sign similarity

ATC transmitted a single communication to an aircraft with a similar sounding call sign (e.g., [Company] 16, vs. [Company] 60) that, if responded to, demonstrated the potential negative effects of partyline generated confusion or improper situational assessment. All five voice condition crews responded to the clearance as though it was intended for them. Three integrated DL crews and four retrofit crews were aware of the transmission.

Turbulence

In previous studies, pilots cited turbulence reports as useful, but crew awareness in this category was highly variable. ATC transmitted three different turbulence reports for altitudes and airspace around our subject airplane during the cruise phase.

The integrated DL condition showed the highest overall awareness: Five crews responded to the first event, four to the second event, and three to the third event. The retrofit condition showed uniformly low awareness; only one crew responded to each event. Crews in the voice condition showed a greater awareness rate than the retrofit crews, but consistently less awareness than the integrated DL crews: Three crews responded to the first event and two crew responded to the second and third events.

Weather

A Lear jet at 51,000 feet reported major storm. A subsequent DL broadcast advisory message described a major storm front from the Canadian border to the Texas panhandle for the three conditions. Awareness ratings were made for the storm report prior to the broadcast advisory. This partyline event rated significant awareness across conditions: four voice condition crews, all five retrofit crews, and three integrated DL crews demonstrated awareness. The weather category captured more crew attention than some other information categories. However, this major storm went unnoticed by three of the fifteen crews. One voice crew remained unaware of the storm. The subsequent DL storm advisory informed all DL crews and demonstrated DL as a viable partyline substitute.

Route information/handling

Flight crews have cited this information category in subjective reports as useful in planning and scheduling activities, especially as they entered terminal areas. The PL indicated heavy traffic activity in the landing pattern after diverting to Denver. ATC issued reduced speed clearances to three aircraft ahead of our subject crews based on this traffic. Awareness in this case could have helped our crews predict the upcoming clearance and plan accordingly.

All conditions showed equal overall awareness to these communications, although the distribution of awareness indicators differed by condition. Three crews from each group registered awareness of the upcoming speed restriction. Voice crews responded to the first such transmission but to no others. It might be argued that, having heard the first transmission, subsequent confirmatory events added no significant new information. The two remaining voice crews failed to register awareness to any of the information. Two integrated DL crews responded after the second transmission, and a third crew responded after the final transmission. Crews in the retrofit condition registered awareness at a rate of one crew per transmission.

An aircraft landing ahead of the crews reported a go-around on final approach into Denver after a windshear attributed 25 knot. airspeed loss. Figure 1 shows that all voice and integrated DL crews heard and responded to this transmission. Three of the retrofit crews indicated awareness. All crews that did not declare a go-around for windshear of their own accord by 1000 feet received ATC go-around orders .

Conclusions

The presence of relevant PL on radio frequencies does not ensure either information detection or understanding of its relevance in a communication environment where crews presumably monitor transmissions for their callsign. Many conditions distract crew attention during relevant broadcasts: approach briefings; navigation/chart activities; cabin attendant communications; system failure management; and crew discussions. Awareness of relevant PL does not necessarily generate crew action. For example, all aware crews continued their take-off after hearing another airplane cleared to cross the active runway. Only one crew requested additional position information from ATC before proceeding. Conversely, windshear and runway traffic conflict events provide extra plan and response time for crews who hear the transmissions, and are clearly of value. Crews without partyline transmissions were, however, able to make timely responses to subsequent (if unanticipated) ATC directives.

The final resolution to this issue is undefined. Technological advances may allow DL to provide information such as turbulence activity, weather reports, and extended approach landing information that assures more certain attention and action than the current system. Other means may provide conflicting traffic and go-around/windshear information (e.g., onboard sensors, ground based radar, or weather sensing equipment). Current results underscore PL detection and response variability in the voice communications environment. The integrated DL condition has no indicated disadvantage compared to voice. Our data is insufficient to show conclusive differences between DL implementation conditions, although crews in the retrofit condition show a numerical disadvantage (out of sixty opportunities for PL detection, the voice condition crews detected 31 events, integrated DL crews detected 32 events, and the retrofit crews detected 23 events). We conclude no evidence found in this study suggests that an early DL environment poses a significant threat to situational awareness due to lack of PL or monitoring.

References

Infield, S. E., Corker, K., Palen, L., Lozito, S., and Hofer, E. (1994) *The Effects of ATC Datalink on Instrument and Environmental Scanning During Flight Operations.* 21st Conference of Western European Association for Aviation Psychology, Trinity College, Dublin, Ireland

Midkiff, Alan H. and Hansman, J. (1992) *Identification of Important "Party line" Information Elements and Implications for Situational Awareness in the Datalink Environment.* SAE AEROTECH Conference and Exposition, Anaheim, CA USA

9 Flight demonstration of data link in an integrated airborne system

Dr G. Richards, DRA Bedford

Introduction

The first flight demonstration of the Mode S data link took place at DRA Bedford in October '91, (see references 1 and 2). The aircraft used was the BAC 1-11, belonging to DRA Bedford and the Mode S ground station was at DRA Malvern. The flight demonstration to be described in this paper is a development of the initial demonstration, using a scenario designed to illustrate a variety of applications of the data link. DRA Bedford devised their own ground station from which ATC messages could be uplinked and downlinked messages received and displayed. ATC control messages were mainly strategic and therefore relatively independent of the latency in the data link system.

Aircraft Systems

The aircraft end system was greatly improved compared with the previous demonstration in that the LCD panel had been fitted with a touchscreen overlay so that push buttons were not necessary, the screen was more effectively used, and integration of the data link with other aircraft systems was greatly increased. The cockpit layout, including equipment used in the data link demonstration, is shown in Figure 1. At the top of the figure, in front of the windscreen, is the Digital Autopilot Control Panel from which the LCD panel with touchscreen was suspended. (The LCD panel could be instantly folded away, if necessary.) The panel consisted of the main screen

FIGURE 1

FIGURE 2

enclosed by a bezel containing 16 touch buttons. On the left side of the cockpit the EFIS display can be seen, consisting of two CRT screens. On the centre pedestal the navigation CDU (RNS 5000) is shown positioned forward on the left side, the transponder CDU is on the right side displaying squawk 3767 and the ARINC printer is at the rear.

Pilot interaction with the data link was made via the LCD touchscreen. All uplinked messages were output on the cockpit printer and strategic data link messages were decoded and displayed on the EFIS map display. The pilot interaction with the data link was as follows. The LCD panel was divided into boxes as shown in Figure 2. All new messages arrived in the top right box, one at a time. Their arrival was indicated on the map display by the flashing symbol 'DL', (at the top right of the map as shown in Figure 2). The pilot acknowledged the message by touching that message on the screen. The message would then disappear from the 'new message' box and, if appropriate, be rewritten in the relevant box below. The flashing 'DL' symbol would also then disappear and if the message was a strategic one it would appear on the map display, using appropriate symbology.

EFIS Map Display

The EFIS map display is the electronic map in the cockpit. Its primary function is to display lateral route information to the pilot. It was also used during this demonstration to display uplink message alerts and strategic messages in graphical form, some of which are shown in Figure 2. A sample of uplinked messages are given in Appendix A.

If the uplink message was strategic in nature, informing the pilot that a manoeuvre had been cleared for action further down the route, orange coloured symbology was placed on the EFIS map at the location of the required action. All messages consisted of an arc at the clearance position, an arrow indicating whether to increase or decrease the parameter or a left/right arrow to indicate the direction of offset required. Each arrow was accompanied by a numeric indication of the clearance. For example ↑260 kts would indicate a clearance to accelerate to 260 knots IAS, ↓FL100 would indicate a clearance to descend to FL100 feet and 4← would indicate an instruction to offset 4 nm to the left of the current track. As the aircraft approached the clearance the numeric portion was made to flash in order to gain pilot attention. This flashing was set to start at approximately 3 nm prior to the clearance to remind the pilot to adjust the autopilot to the correct settings before engaging the new settings overhead the required position.

If the autopilot was engaged early the alert would continue flashing until overhead the actioning position, to inform the pilot that the change had not yet been cleared. If the autopilot was engaged late or with incorrect settings then the alert continued flashing until the error had been corrected. The other indication which appeared on the EFIS map was associated with the route clearance. The portion of the route cleared by the ground system was coloured white, the portion not yet cleared was coloured orange, clearance message.

Volmets

The VOLMET service already supplied by NATS is a series of VHF radio broadcasts which each cover approximately seven airfields. The pilot is required to wait until the airfield of interest is announced and then hastily copy down the information spoken to him. Often the information is not easily discernible and the pilot is required to wait for the airfield to be announced again. This process may be executed for both the destination airfield and at least one diversion. This can be a lengthy process requiring the pilot's undivided attention for many minutes at a time.

The data link VOLMET service has many advantages over the VHF version now that the database has been digitised. During the demonstration, using the data link CDU, the pilot was presented with a list of 27 airfields covered by the service, identified by their standard ICAO codes. These were presented on three pages, as shown in Figure 2, between which he could move freely. VOLMET information for the required airfield was obtained merely by the pilot touching the appropriate airfield code. The request was formulated and downlinked enabling the required information to be extracted from the ground database. The ground database is updated regularly with data from the Met Office. The information was formulated as a coded uplink message and was delivered to the aircraft. The VOLMET was presented to the pilot both on the CDU and on the cockpit printer. The printout provided a permanent copy leaving the pilot free to request another VOLMET as required, which overwrites the previous VOLMET on the CDU. Each VOLMET request takes, on average, between 12 and 18 seconds to complete, the main delay being the 6 second aerial rotation period.

Summary

The four aspects of the data link demonstrated were:

-        uplinked and downlined ATC messages

- data link display integrated into the cockpit
- VOLMET request available from 27 airfields
- downlinked extracted avionic data.

All data link information was displayed on an LCD panel, centrally placed in the cockpit. Pilot interaction with the data link was achieved by use of a touchdscren overlay on this LCD panel. The pilot found this display easy to use although it was positioned more ideally for demonstration viewing. Uplinked messages were mainly strategic, for which the data link is well suited. (Tactical messages should be preferably passed over RT). Strategic messages were also displayed on the EFIS map display which the pilots found particularly useful.

On the previous data link demonstration, VOLMET obtained over the data link proved to be a vast improvement over listening to the VHF transmission of VOLMETs for a series of airfields. It was shown again in this demonstration how the VOLMET for a particular airfield could be requested and a hard copy obtained. This was achieved by paging through a list of 27 airfields on the LCD panel, arranged in alphabetical order, and simply touching the airfield name required.

References

1.   *A Report on Flight Trials to demonstrate a Mode S data link in an ATS environment* - Eurocontrol UKCAA, DRA

2.   *A Report on Flight Trials to Develop, Test and Demonstrate an Experimental Mode S Data Link in an ATS Environment* - G Richards.

Appendix A
Sample Uplink Messages

Please downlink performance parameters.
Route received by ground system.
Maintain FL070 and 260 kts.
Route cleared to RNB.
10 nm after CTM climb FL200.
10 nm before CSL offset LEFT
  4 nm until RNH.
After WTM add waypoint CAB.

At RNA accelerate 260 kts.
Please downlink your route.
Delete waypoint SKT.
10 nm before RND climb
  FL240 accelerate 270 kts.
Contact SECTOR 1 on
  379.25.

10 The effects of ATC datalink on instrument and environmental scanning during flight operations

Susan E. Infield, Leysia Palen, David Pepitone, Steven Kimball and David Smith, Boeing Commercial Airplane Group
Antonio Possolo, Boeing Computer Services
Kevin Corker and Sandy Lozito, NASA Ames Research Center

A Datalink (DL) research study assessed the effects of two DL implementations on head-down and head-away times in a full mission simulation. The study used a 747-400 engineering simulator with full visual simulation capability. Crews flew in one of three conditions: voice communications, flight management system (FMS) integrated DL, or retrofit DL. Researchers collected eyetrack and head-movement data with lightweight, helmet mounted equipment. This paper discusses the evidence of DL impact on both environmental and instrument scanning activities.

Operational issues

The introduction of two way DL communication between aircraft and air traffic control (ATC) and between aircrews and their parent companies raised several human factors issues. One issue generated from the ATA document *Human Factors Requirements for Datalink Systems Implementation* and from pilot's concerns is the effect of DL on system and environmental monitoring. Flight crew members subjectively report that they remain free to scan the external environment and their primary flight instruments during voice communication with ATC and ground support facilities. At least one crew member is required to access, read, and communicate messages displayed on a flight deck interface device in a DL environment. The time to complete these actions and initiate ATC clearances must be considered in the total context of flight crew operations.

Method

Subjects We recruited fifteen current and qualified line crews from two major U.S. domestic airlines. Crew members flew in their customary positions (captain or first officer). These volunteer crews received no financial compensation for their participation beyond hotel accommodations and food during their stay in Seattle.

Design The experiment used a between groups design with two independent variables. All crews flew in a single condition but experienced identical routes with the same set of clearances, enroute traffic, weather, and mechanical failures. Communication mode was the principal manipulation with three conditions for all clearances and company communications: (i) voice communication - used normal voice radio channels; (ii) FMS-integrated DL - used a two-way DL implementation on the forward control display unit (CDU). This implementation interfaced with the FMS and permitted direct loading of ATC directives into the system; and (iii) retrofit, non-FMC-integrated DL - used a two-way DL implementation which used the center CDU. This implementation did not interface with the FMS or permit direct loading of ATC directives into the system. A second manipulation, discussed in a separate paper (Infield, Corker et. al, 1994), involved "partyline" manipulation, i.e., information overheard from conversations between ATC and other nearby aircraft. The voice communication condition had a full set of scripted partyline. The two DL conditions had a reduced amount of overall partyline. This simulated an early environment with approximately 25% DL equipped commercial aircraft. The ratio of relevant to irrelevant partyline was the same for all communication conditions. The dependent variables of interest to this paper were dwell times and scan patterns for specific areas of interest (AOI). Our analysis specified these AOIs: 1ST and 2ND EICAS; captain's (CAPT) CDU, ND, and PFD; first officer's (F/O) CDU, ND, PFD; MCP; PEDESTAL; WINDOW; and OFF. We introduced a derivative AOI PEDESTAL, comprised of the CNTR CDU and F/O RADIO, to alleviate possible erroneous apportionment of dwell times between these two AOIs. OFF captured dwell times and saccades on anything other than defined AOIs.

Scenario The detailed flight scenario contained ground and airborne traffic conflicts, weather advisories, and mechanical faults. Flights originated at San Francisco International bound for Washington National. High fidelity, out the window visual scenes encouraged environmental scanning. Subjects encountered other aircraft during initial taxi and in flight. Inflight events included routine traffic advisories and one emergency traffic evasion. The detailed scripts included all clearances to the subject aircraft and all partyline conversations between ATC and traffic during the flight We translated voice clearance scripts into DL format for the two DL conditions. A highly experienced, retired controller provided ATC support. Experimental personnel (all qualified pilots) provided traffic conversation. The scenario contained a mechanical fault which resulted in a diversion to Denver after an average of 140 minutes. Average times for individual flight phases are as follows: Take-Off – Cruise: 16 minutes; Cruise – Descent: 87 minutes,; Descent – Go-Around: 18 minutes; Go-Around – End: 12 minutes.

Equipment We used a 747-400 engineering simulator with single channel-forward view visuals and systems capability. Researchers constructed two DL implementations. One system, integrated into the FMS, used the forward CDUs, (both FO and CAPT side) as the message display and composition device. Crews could directly load some clearance values (most notably route changes) into the FMS with the LOAD and EXECUTE CDU buttons. A second system shared general design principles and functionality with customer proposed retrofit DL implementations. This DL system used the center CDU at rear of the pedestal as the interface device. This system was not integrated into the FMS and required manually input clearance values, as with voice communication. We formatted messages according to SC169 MOPS requirements in both systems. Both systems allowed crews to compose clearance requests and free text messages for downlink to ATC or company representatives. An ASL helmet mounted eyetrack data collection device recorded both eye and head movement data. Low light IR video cameras mounted behind the crew and in the front corners of the simulator recorded video and audio activities. Data streams from the simulator recorded aircraft movement, all switch positions, crew input, system state and timing data.

Procedures Researchers randomly assigned subject crews to experimental conditions with the reservation that equipment problems in either of the DL conditions might re-assign a crew to an alternate condition. Crews received a briefing prior to the experiment, completed demographic and experience questionnaires, received a cover story, and remained unaware of the full intent of the experiment. Researchers assigned the F/Os principal responsibility for communications activities, although CAPTs retained ultimate authority to assign tasks. F/Os received training in one of the DL implementations. Training averaged about 35 minutes and directly preceded the experiment. CAPTs received similar briefings, but were not required to reach the same level of competence as F/Os. Researchers fitted the F/Os with helmet mounted eye tracking equipment, calibrated and tested the equipment, and supplied crews with final dispatch papers. Data collection began with the crews' preflight briefing and ended after touchdown in Denver.

Results

Data Validation Analysts computed and compared three different estimates of flight duration to validate the head-tracker data for each flight: (i) total fixation time plus total saccades time; (ii) total recorded clock time (tracker clock); and (iii) total flight duration (take-off time minus flight end time, according to the video tape recorder clock).

One retrofit DL flight showed a gross deficiency in head-tracker time relative to the actual flight time. We excluded the information gathered during this flight from all further analyses.

Dwell times Dwell times are the total times spent fixated onto that AOI plus the duration of any saccadic gaps that occurred between successive fixations onto that AOI. Both fixations and saccadic gaps were determined by the software bundled with the head-tracker based on assumptions about the spatio-temporal stability of the subject's line of sight.

Dwell times for each flight phase were determined and statistics were derived from them: in particular, several re-expressions of the dwell times themselves, and head-down and head-away times.

Interface use time Analysts compared DL implementation impact on interface device interaction to voice communications. They applied Wilcoxon rank-sum tests to the F/O CDU dwell percentages for voice and integrated DL; and to the PEDESTAL dwell percentages for voice and retrofit DL.

There is a significant difference (rank-sum statistic W=16, n=5, m=5, p-value=0.02) between voice and integrated DL during GoAround–End flight phase.

The p-value is 0.095 between voice and integrated DL during Cruise–Descent flight phase.

When we compared voice with retrofit DL for dwell percentages onto PEDESTAL, we found significant differences during Takeoff–Cruise (rank-sum statistic W = 70, n = 5, m = 12, p-value = 0.006) and Cruise–Descent (rank-sum statistic W=72, n=5, m=12, p-value = 0.002) flight phases.

The p-value is 0.08 between Voice and DL (Rear) during the Descent–GoAround flight phase.

Analysis showed greater DL interface device use time in the DL implementation than in voice. The comparison of voice and DL implementation for other flight phases did not approach significance.

We carried out similar tests to compare dwell times onto the two different DL implementations, for each flight phase. The results, summarized in the next table, suggested no significant differences.

Flight Phase	W	m	n	p-value
TakeOff–Cruise	24	5	4	0.90
Cruise–Descent	27	5	4	0.73
Descent–GoAround	26	5	4	0.90
GoAround–End	33	5	4	0.064

Head-down time Head-down is the total dwell time spent focused on non-WINDOW AOIs. This is actual time, *not* a percentage of the total dwell time onto all AOIs. Also, head-down includes fixations onto OFF.

Analyses of variance, carried out for the data pertaining to each of the four flight phases, showed no significant differences between the three communication modes in any flight phases.

The following table shows results for the TakeOff to Cruise flight phase.

Heads-Down ANOVA: TakeOff to Cruise					
Source	Df	Sum of Sq	Mean Sq	F Value	Pr(F)
Comm. Mode	2	308.321	154.1607	0.972	0.41
Residuals	11	1744.152	158.5593		

Total head-down dwell time proportions showed no significant differences.

We also found no significant differences when we disregarded dwell times onto OFF.

66

Head-away time Head-away time is time deducted from primary flight instruments monitoring other than WINDOW. Head-away is defined separately for each of the two DL implementations .

We decomposed dwell times for specific primary instruments and compared them across the three conditions to explore relative differences in instrument monitoring times. For integrated DL, heads-away time is the percentage of total non-F/O CDU dwell time for each AOI other than F/O CDU. In the retrofit DL, head-away time is the percentage of total non-PEDESTAL dwell time for each AOI other than PEDESTAL. These are renormalized percentages and add up to 100%, exclusive of WINDOW or OFF time. We computed these percentages separately for each flight phase.

We added the renormalized proportions of head-away over AOIs for each flight phase for each flight (other than OFF and WINDOW, which are removed via renormalization) and obtained samples for each DL implementation. The following Wilcoxon's test results comparison suggest no significant differences:

Flight Phase	*W*	*n*	*m*	*p-value*
TakeOff–Cruise	26	5	4	0.90
Cruise–Descent	21	5	4	0.41
Descent–GoAround	22	5	4	0.56
GoAround–End	17	5	4	0.064

We compared each DL implementation to voice with the following results:

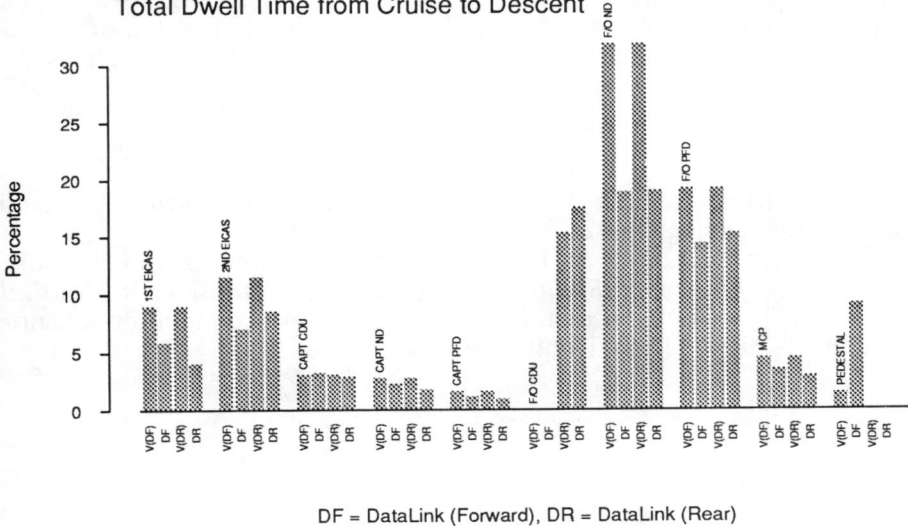

DF = DataLink (Forward), DR = DataLink (Rear)

Figure 1 Datalink vs voice heads-away times

67

Comparison	Phase of Flight	W	n	m	p-value
V-DF	TakeOff–Cruise	31	5	5	0.55
V-DR	TakeOff–Cruise	35	5	4	0.016
V-DF	Cruise–Descent	38	5	5	0.032
V-DR	Cruise–Descent	35	5	4	0.016
V-DF	Descent–GoAround	39	5	5	0.016
V-DR	Descent–GoAround	35	5	4	0.016
V-DF	GoAround–End	37	5	5	0.056
V-DR	GoAround–End	35	5	4	0.016

Conclusions

The time dedicated to DL operations significantly increases the amount of DL interface device interaction time. The question remains, how is the additional time acquired? With the scope of our experimental design, we consider three possibilities. The time increase is derived from: (i) reduced environmental scanning (i.e., increased head-down); (ii) reduced instrument scanning; (iii) non-essential scanning (OFF). This report's data indicates no compelling evidence to support reduced environmental scanning concerns. The low subject size and the high degree of inter-subject variability certainly reduces our power to detect differences. Therefore, treat this finding as optimistic (though hardly conclusive). Our data yields contrasting evidence of reduced system scanning in the DL environment in virtually all flight phases. This effect is not present for integrated DL during the Take-off phase or following the Go-Around event. Whether these statistically significant findings translate into operationally meaningful values, the possible mitigating effects of increased procedural training, and the effects of practice in reapportioning workload are questions which must now be addressed.

References

Human Factors Requirements for Data Link: ATA Information Transfer Subcommittee document, 1992

Infield, S. E., Corker, K., Logan, A., Lozitio, S., Palen, L., Hofer, E., and Possolo, A. (1994) *The Effects of Reduced Partyline Information in a Datalink Environment.* 21st Conference of Western European Association for Aviation Psychology, Trinity College, Dublin, Ireland

11 Flight crew performance in automated air traffic management

Kevin Corker, NASA Ames Research Center
Sandy Lozito, San Jose State University
Greg Pisanich, Sterling Software Inc, NASA Ames Research
Center, Moffett Field, California

Driven by concerns for safety in an increasingly crowded airspace and by concerns for economic concerns NASA has recently initiated a research program to enhance Terminal Area Productivity (TAP). This initiative seeks to focus technologies in automated air traffic management, improve flight management systems, and improve operations in low-visibility landing and surface operations. In addition to aiding systems for the controller and flight deck operations, efficiencies in the medium of air/ground information transmission are being explored in digital communication (data-link) operations. As these technologies develop, we have undertaken empirical studies in full-mission simulation to assess the impact of these advances on flight crews. Fuel-efficient automation in flight management must be operable in the automation-aided control environment in order to realize the promise offered by technologies such as the Center TRACON Automation System (CTAS). The goals of efficient flight operations favor the use of automation to the extend possible. Additionally, the rapidly developing capabilities of automated aiding technologies for both ground control and flight deck have created a requirement for an effective and efficient predictive methodology to examine the impact of variations in systems. This need is being met through the development of a human performance model of flight crew operation that includes predictive mechanisms for memory, decision-making, performance time and information sufficiency. This model as applied to CTAS clearance delivery will be described. It is the purpose of the model to predict the likelihood of the effective use of automation in an automated air traffic management environment. In support of this modeling effort, data from the Air/Ground Compatibility study were used to support model development and verify model operation.

Introduction

The goal of the Terminal Area Productivity (TAP) Program is to provide technology and operating procedures for safely achieving clear-weather capacity in instrument-weather conditions. The challenge presented by the TAP program goal is one that involves linking human operators, computer aiding systems, and complex flight management systems in a tightly coupled, dynamic, closed-loop system. In order to successfully meet the program goals, we have undertaken a series of studies that examine the role of the human operator (both flight crew and air traffic controllers) in the advanced air traffic management system.

Automated ATC/Flight Deck Integration

Integrating terminal-airspace control-aiding systems, specifically CTAS, with flight deck flight management systems is critical to achieving accurate and predictable aircraft arrival operations (Erzberger, Davis, and Green, 1993). The flight management systems in modern "glass cockpit" aircraft are designed to support effective fuel efficient operation of the aircraft in a range of flight regimes. The CTAS has as its goal optimal sequencing and scheduling arrival traffic into a terminal airspace. CTAS provides clearance advisories to the air traffic controller that can specify the descent profile and point of initiation of descent for an aircraft from cruise phase operations (top of descent or TOD). The clearances can indicate descent speed and profile including a specification of time of arrival at waypoints, essentially a 4-D clearance with latitude, longitude, altitude and time provided. The flight crew is currently provided this clearance information by voice radio transmission; although digital transmission of information between the ground controller and the flight deck (data-link) is a near term technology advancement.

Air/Ground Compatibility Study

A simulation experiment was conducted using a generic glass cockpit simulator with full motion and visual capabilities. The simulation was designed as a line-oriented-flight-training scenario originating in San Francisco enroute to Denver Stapleton Airport where the subject aircraft was directed in descent and approach via the CTAS system. The experimental flight was flown by ten current line-operational flight crews from a single U.S. air carrier.

The two major manipulations in this experiment were datalink v. voice clearance transmission (the datalink implementation used the FMS/CDU as the interface), and conventional v. CTAS airspace control in the Denver Terminal Area. For the purposes of this paper, the data we have examined deal with crew procedures around the top-of-descent point in both voice and data-link communications as a function of CTAS-generated clearances.

Predictive Human Performance Model

Effective human-centered design is dependent on adequate models of human/system performance in which representations of the equipment, the human operator(s), and the mission tasks are available to designers for manipulation and modification.The joint Army-NASA Aircrew/Aircraft Integration (A^3I) Program, with its attendant Man-machine Integration Design and Analysis System (MIDAS), is an example of this type of predictive model development. MIDAS provides designers with a test bed for analyzing human-system integration in an environment in which both cognitive human function and "intelligent" machine function are described in similar terms. The focus of this modeling is characterization of the process of perception, decision-making, activity selection, timing, and task-loading incurred by the operator as he or she interacts with the system under study. These models describe (within their limits of accuracy) the responses that can be expected of human operators in several areas that are critical to safe and reliable operation of advanced automated systems. The full MIDAS system description can be found in Corker and Smith (1993).

Simulation and Experiment

Method

Air-MIDAS was developed using the MIDAS simulation architecture and editing tools. The challenge was to explore an optimal range of time for the issuance of a CTAS clearance for descent so that the aircrew is likely to accept the clearance and enact it with fight deck automation. CTAS calculations tend to favor a late issuance (close to the TOD); while the flight crew tends to favor clearance information as early as possible to aid them in planning and configuring the aircraft.

The basic simulation driver is time, using an increment length of 100 milliseconds. All simulation events (e.g., time of clearance issuance, interruptive activities, traffic events, and communication interrupts) were generated using Markov probability distributions tailored to represent stochastic models developed. These distributions were generated using expert opinion and human performance data from the air/ground compatibility experiment. Human performance data were generated from analysis of the video recordings of performance in the manned simulation. These data were verified and resolved through computer output data of each flight in the experiment.

Activities

A sequential activity set was developed for the aircrew tasks associated with servicing the top of descent CTAS clearance. In the case of a nominal clearance those activities are to be performed as follows: Receive CTAS clearance from ATC; Determine if time is available to decide on the clearance; Determine whether the descent can safely be initiated; Communicate intent to accept or reject the clearance to ATC; Determine automation level for the descent procedure; Implement descent procedure.

Using this model as a framework, the associated leaf-level activities, decision rules, and information requirements were investigated for each communication/automation combination. These were encoded using the MIDAS activity description language. For example, the leaf-level activities associated with receiving a clearance by datalink included such activities as viewing the clearance and accepting the clearance. The completion times of these activities are represented as stochastic distributions of the performance exhibited by the air/ground compatibility experiment air crews. Interrupts are a second class of activities represented within Air-MIDAS which do not contribute directly to the completion of the clearance task (such as responding to traffic calls). These activities can suspend the current clearance activity, delaying its completion while the interrupt is serviced. The following interrupt activities are represented: TCAS traffic advisory (TA); Visual traffic call by crew; ATC messages to other aircraft; ATC message to own aircraft; ACARS message; Request to repeat message by crew member; Cabin crew request.

An interrupt resumption method is also assigned to all leaf level activities. If interrupted, these methods describe how the activity is resumed, which can affect the time required to complete the task. Methods implemented include: non-interruptable, resume, resume-from, and restart.

Decision Rules

The rules required to make decisions based on time available, safe descent, and automation were derived from expert opinion. For example, the rules associated with safe descents include weather, equipment-limitations, and passenger comfort considerations.

Updatable World Representation

The UWR structure in Air-MIDAS represents the information that the aircrew understands about the world. Categories of UWR information include the aircraft state and configuration, meteorology, descent information, traffic, and pilot operational knowledge.

Simulation and Experiment Results

In order to have comparative data between the simulation study and the MIDAS model, timing data between the CTAS clearance delivery and the aircraft's top of descent were collected (See Figure 1).

Crew / Condition	Message Receipt (DL) Message Beginning (voice)	Message Access (DL) Message End (voice)	Message Accept (DL) Readback Complete (voice)	Top of Descent Point Reached	Method of Flying Descent
19/Voice	6:48:49	6:48:59	6:49:21	7:07:58	FLCH
20/Voice	7:02:10	7:02:18	7:02:24	7:07:32	FLCH
21/Voice	7:00:22	7:00:31	7:00:38	7:07:34	V-NAV
22/DL	6:53:56	6:54:03	6:54:18	6:58:32	V-NAV
23/DL	6:55:23	6:55:26	6:55:42	7:00:34	V-NAV

Figure 1. Times and method of flight associated with crew behaviors between the CTAS clearance transmission and the aircraft's top of descent (TOD) time. These are the tasks relevant to the crew's meeting the CTAS clearance requirements within that time period.

Model Runs

As a preliminary examination of the Air-MIDAS simulation, a 3 X 2 X 5 full factorial experiment was run with 100 replicates for each factor-level combination. (Illustrated in Figure 2.) The factors manipulated were: automation mode (autoload, FMS, and MCP), communications mode (datalink, and voice) and time provided to implement the clearance (ten times separated by 5 second intervals). The weather and altitude, and calculated descent rate were allowed to vary within a reasonable range. Whether or not the aircrew was able to complete the clearance successfully within the time given was the dependent variable which we labeled "success".

Discussion

In examining the data output from the MIDAS model, we see that as the TOD point is approached the aircrew model decides to select a flight mode that involves less automation (MCP operations). Further, the time/distance from TOD that the choice to switch from more to less automated operation is sensitive to the medium through which the clearance is presented with voice presentation of the data providing a more pronounced use of non-automated modes. This mode switch interaction with medium of exchange is also reflected in the human performance data provided.

The model "behaves" in a way that is consistent with human crew operation in the mode selection process. The model also confirms the hypothesis that as the TOD point is closer the aircrew will select the lesser automated alternative mode of control. Finally, as the TOD point is approached around 8 to 5 miles from the CTAS required TOD point, the number of successes in any clearance compliance is reduced.

Conclusion

The effort to produce a model that predicts human flight crew behavior was undertaken to provide an analytic tool to answer questions about human/automation interaction in advanced airspace management. The preliminary results indicate that the MIDAS model structure will serve the purpose of analysis. The initial success with limited parameters suggests that the model parameters be further manipulated and the model coverage be expanded to other flight phases, and that specific verification experiments be performed to establish model performance.

References

Corker, K. M, and B. Smith (1993), 'An Architecture and Model for Cognitive Engineering Simulation Analysis', Proceedings of the AIAA Computing in Aerospace Conference, October, San Diego, Ca.

Erzberger, H., Davis, T. and Green, S. (1993), 'Design of Center- TRACON Automation System', AGARD Guidance and Control Symposium on Machine Intelligence in Air Traffic Management,May, Berlin, Germany.

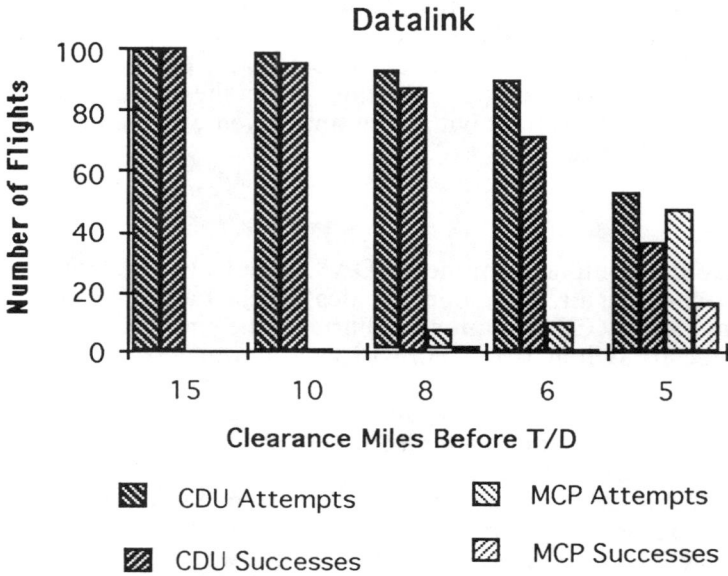

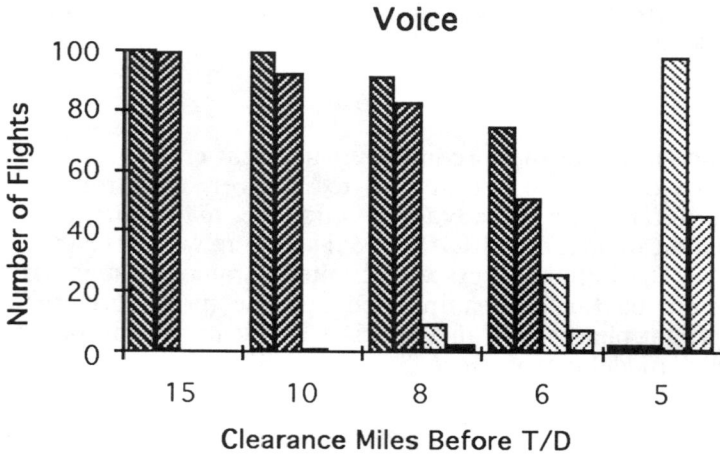

Figure 2. Flight control mode selection as a function of distance to TOD and medium of clearnace delivery.

12 Input of ATM on future flight deck design

P.G.A.M. Jorna, National Aerospace Laboratory, Amsterdam, The Netherlands

Introduction

Air traffic management concepts aim at increasing air space capacity and throughput by more accurate planning, navigation and the use of time based operations, effectively creating a 4D environment. Pilots and controllers are still regarded to be critical components of the future system and they should be supported by optimized man-machine interfaces to assure adequate performance. Multiple strategies can be identified in accomplishing the challenge. The roles of pilot and the tasks allocated to human operators will change. With more aircraft, safety also needs to be enhanced drastically Both requirements will influence future flight deck design. Two possible cockpit extremes are identified and discussed.

The ATC bottleneck

Present ATC systems are either loaded to their maximum capacity or are approaching such a level. The traditional way of increasing ATC capacity is to open more sectors. The same airspace is divided in smaller sections so more controllers can deal with the aircraft. The communication load of handing off these aircraft to adjacent sectors will, however, rapidly increase as a function of the number of sectors per airspace. A doubling of the sector will therefore only result in an estimated capacity increase of around 30 %.

The controller will guide and vector the arriving traffic on a 'first come first serve' basis. Such tactical control is presently not compatible with the FMS (Flight Management System) capability reducing its benefit for the

75

airline. Re-programming the FMS after an ATC request is cumbersome, so it is switched off in the flight phase where it should have the greatest advantages, i.e. in the busiest areas with the highest task loads for the pilots.

Future ATM strategies

Reduction of uncertainties is essential to meet the anticipated demands. A more accurate and better updated aircraft position is helpful in reducing separation. Knowing the intentions of aircraft will reduce the number of potential course changes to consider in conflict resolution etc. Better planning could reduce the number of interventions issued by ATC and evaluating aircraft generated trajectory proposals would allow the capability of the FMS to be exploited more fully. All ATM strategies include a time based component to enable the aircraft to fly so called 4D trajectories (EUROCONTROL, Pozesky 1991). A digital datalink of sufficient capacity is needed to enable effective data exchange between aircraft and ground systems.

The economical perspective of the airlines dictates them to strive for as much room for flexibility as possible. Airlines compete on issues important to their passengers, i.e. comfort, short flight times, special destinations all over the world and affordable ticket prices. They are therefore expected to strive for aircraft with low interdependencies on ground services.

Alternatively, the ATC ground systems are regulated and have to cope with multiple parties and their job will be easier if everybody simply complies with instructions and assigned trajectories.

Performance requirements will affect the allocation of functions to be either executed as a human task or by a piece of automation.

Some will argue, with due reason, that the human is intrinsically unreliable and will opt for a full automation solution. Others will argue that it is impossible to assign legal responsibilities to this piece of software and so the human is kept as a backup in the system. Both perspectives lead to jokes like the "future pilot with his dog" that will to bite the pilot, if he or she touches anything. Clearly, if trust in human performance, endurance and reliability is to be restored, some major change in tools should be accomplished.

Man-Machine configurations

ATM concepts take different views on the role of human operators. As a consequence, Man -machine interface requirements and research will therefore also differ. An evolutionary approach, as an example, would dictate a high commonality with existing systems, while a major change in performance and safety level would dictate a more radical experimental concept to pursue.

Some related areas in aeronautical research for ATM are summarized in figure 1.

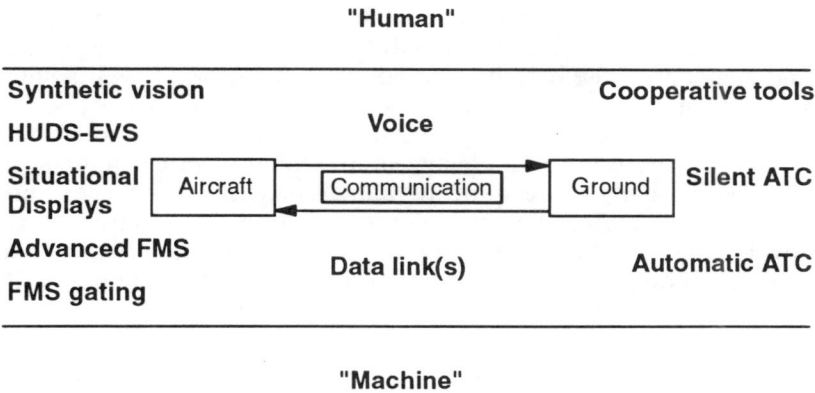

"Human"

Synthetic vision			**Cooperative tools**
HUDS-EVS		**Voice**	
Situational Displays	Aircraft → Communication → Ground		**Silent ATC**
Advanced FMS		**Data link(s)**	**Automatic ATC**
FMS gating			

"Machine"

Fig. 1 Aeronautical Research

On the aircraft side, isolated research activities are focussed on 'head down' systems like the FMS or on 'head up' systems like enhanced vision through integrated sensors and HUD's. One strive is for more compatibility between ATC and FMS systems. Data links will allow data to be exchanged, provide support for a negotiation process at different levels of authority and enable the issuing of complex 4D clearances. More advanced FMS functions depend on an extensive integration with Data link. Human operators in that case, could serve almost as a nuisance factor as they will absorb unpredictable time periods, for evaluating, discussing and executing the instructions, thereby exacerbating the time delays already involved in digital data transfer by mode S radars and satellite communications (Brüggen 1993).

On the ground side, there is the strive for providing controllers with tools that organise their data in useful, task related information. There is, however, at the same time a technology driven development for fully automated ATC functions like the ARC 2000 at Eurocontrol or the NLR datalink equipped CTAS version (Center TRACON Automation System) developed in collaboration with NASA AMES.
Economic pressures could initiate an ad-hoc development of future flight decks without sufficient validation of human performance levels or work strategies when using the new equipment. Design and validation will have to go hand in hand as there is no comparable research data base to rely on. So, new knowledge will have to be gathered in experiments.

Glass cockpit: some lessons learned

Understanding and predicting the behaviour of cockpit automation is a general concern, as well as the potential erosion of flying skills. The cockpit itself is appreciated by many pilots with a particular emphasis for the EFIS (Electronic Flight Information System). The map display allows a compelling overview of position and route and is therefore commended for a positive effect on 'situational awareness'(Dorp van, 1991). Overreliance and overconfidence are, however, areas of concern. The clarity of the display (independent of the data it is based on) is a factor, as well as the reliability of the automation which can induce so called 'complacency'.

Many modes exist that considerably alter the behaviour of the aircraft but are only known to the pilot by a change in an alphanumerical indicator. Unexpected mode changes are quite regular (Corwin, 1993 Sarter and Woods 1992) and difficult to detect as the information, or system data, is scattered over different locations. Insufficient feedback or 'observability' is a general cause of complaints. Misinterpretations of aircraft behaviour will increase the potential for fatal accidents with otherwise intact aircraft.

The Glass cockpit of the first generation still presents pilots with all possible data, most often in a glass version of a previous mechanical instrument representation, but it is still the pilot who has to integrate that data to obtain task relevant information.

Feedback from system responses to pilot inputs and intended automation activities are essential for predicting system behaviour and enabling fault management and error correction during data entry or mode selection. Last but not least, the cumbersome handling of some interfaces should be simplified to improve the operability and prevent unnecessary 'head down' times, meaning distractions from the outside or the primary instruments.

Perspectives on future flightdecks

Two major, or extreme, cockpit concepts seem to emerge if one reviews the research initiatives in the field. A concept closest to a further development of the present FMS equipped aircraft is what could be denoted as the DRONE concept (Data link Routed Obedient Navigation Environment). Ground instructions are uplinked and auto-loaded in the FMS for instantenuous execution. Pilots essentially serve in a back-up mode.

An alternative view point is to strive for 'electronic VFR', with the promise of expanding IFR capacity to VFR levels or even beyond. Information has to be made available that increases traffic awareness to such levels that local conflict resolution is possible.

This PRIDE concept (Pilot Routed Informed Decision Environment) intends to exploit the flight crews capability for decision making and initiative, by providing them the information and means to maintain, negotiate and amend their flight trajectories. They serve as active economic components of the system and not mere back-ups for the unexpected.

The main characteristics are summarised in fig. 2.

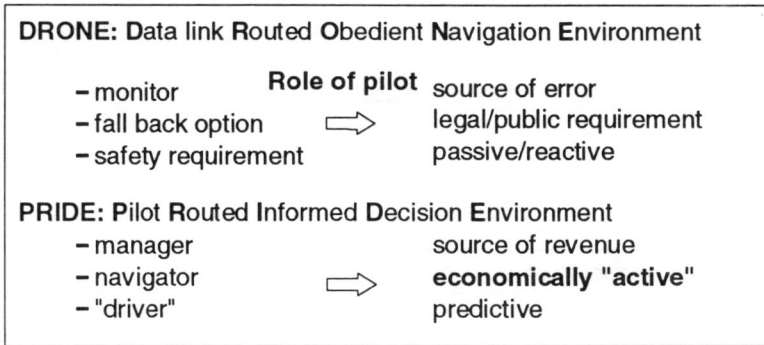

DRONE: Data link Routed Obedient Navigation Environment

　– monitor 　　　**Role of pilot**　source of error
　– fall back option　　　⇒　　legal/public requirement
　– safety requirement　　　　　passive/reactive

PRIDE: Pilot Routed Informed Decision Environment
　– manager　　　　　　　source of revenue
　– navigator　　　　⇒　　**economically "active"**
　– "driver"　　　　　　　predictive

Fig. 2 Cockpit Concepts (Extremes)

An important determinator for actual development will be the level of sophistication of ATC achieved worldwide. Drone aircraft will perform best in a "high Tech ATC" environment, while PRIDE aircraft have the potential to operate effectively in "low Tech ATC" environments. Both concepts clearly represent extremes, but illustrate the different thinking about the reliability and performance capabilities of pilots (and controllers). Removing them from the system will not remove the notorious 'Human Error' but simply change its occurrence to other locations, for instance software engineering or systems analysis.

Research strategy

Research concerning human factors should provide objective empirical information that support a functional design of an aircraft that combines the best of both the DRONE and PRIDE flight deck concepts. Both alternative concepts are subject of investigation in the'Pilot performance in automated cockpits' program at NLR. The studies so far included reviews of cockpits, experiments concerning the informational value of moving throttle levers, pilot effectiveness with potential cockpit data link interfaces, 4D informational requirements and advanced display developments with an emphasis on air/ground compatibility.

The required testfacilities for ATM research are considerable as depicted in fig. 3. Air-Ground compatibility research requires that reconfigurable cockpit systems are tested/validated against different ATC/ATM prototypes.

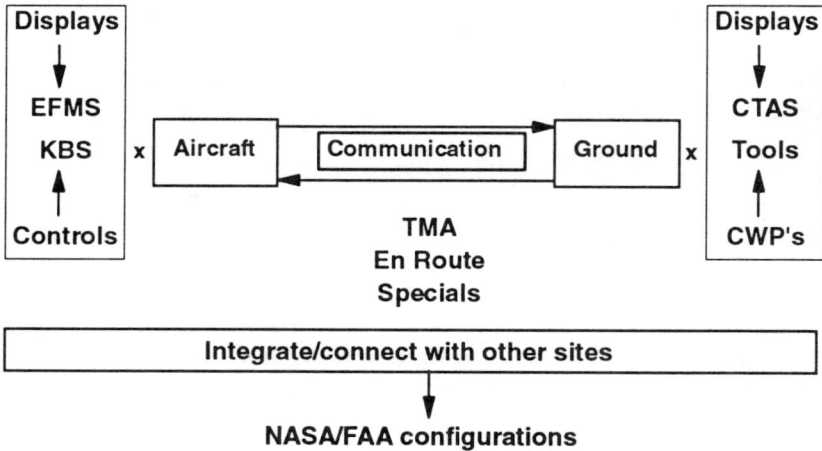

Fig. 3 NLR Human in the loop validation

International collaboration is needed to obtain cost effective and mutually supported concepts and solutions. Recently, research facilities have been networked between NLR, NASA and the FAA. The dutch research simulator can now land in the United States under control of experimental ATC or allow dutch control over traffic in an US terminal area. Potential research flight deck designs can be realised on separate simulators and tested under the same real time scenario. Both DRONE and PRIDE flight decks will (have to) be subject of extensive research in the fields of human factors and ATM effectiveness.

References

Brüggen, J.(1993). *On speaking terms with the EFMS through digital data link*. In proceedings of the PHARE FORUM. October, Braunschweig.

Corwin, W.H.(1993). *Autoflight Mode Annunciation: Complex Codes for Complex Modes*. Seventh Int. Aviation Psychology Symposium. Ohio.

Dorp, van A.L.C.(1991). *Pilot opinions on the use of flight management systems*. NLR TP 91076 L. Amsterdam, the Netherlands

Pozesky, M.T. (1991). *The future Air Traffic Management System for the United States*. Aviation of safety Journal. Vol 1. No 2.

Sarter, N.B. and Woods, D.D.(1992). *Pilot interaction with cockpit automation: Operational experiences with the Flight management System*. International Journal of Aviation Psychology, 2(4),303-321.

Part 3
ATC: HUMAN FACTORS

13 Human Factor design considerations for Air Traffic Control information displays in a modern glass cockpit

Prof. Dr.-Ing. Gerhard Hüttig
Dipl.-Ing. Andreas Hotes
Dipl.-Psych. Andreas Tautz
Berlin University of Technology, Institute of Aeronautics and Astronautics
Section Flight Guidance and Control/Air Transportation

Introduction

To accommodate the future anticipated increase in air traffic, the Air Traffic Control (ATC) systems not only have to be harmonised, but have to provide increased capacity by using enhanced technology on ground and aboard. One approach will be the implementation of air-ground data link, providing the Air Traffic Management (ATM) computers with all flight information derived from aircraft on-board Flight Management Systems (FMS) (Bohr, 1990).

In the future, digital air-ground data link will enable the transmission of automatically generated ATC messages into the cockpit. The feasibility of the visual presentation of such information is studied, using an Airbus A340 full flight training and research simulator with a generic display generation system. The study addresses in particular the development and integration of a display into an advanced glass cockpit, which presents tactical ATC messages. The procedures to accept the incoming ATC messages and the consequences on the visual channel and mental workload of pilots working with such displays are investigated.

Research capabilities

Future ATM concepts and the integration of respective onboard functions require the optimisation of the Human Machine Interface (HMI) in a highly automated aircraft environment. The A340 full flight training and research simulator at the Technical University of Berlin provides respective research capabilities. The implemented Flight Management System Simulation (FMSS) provides a very close replica of the A340 FMS man machine interface. The interface includes the interactions with the Primary Flight and Navigation Displays/ Flight Mode Announciators, the Flight Control Unit, the Multipurpose Control and Display Units and could therefore be used to simulate the air/ground data link functions.

A very important research tool of the simulator's Scientific Research Facility (SRF) is the Experimental Data Unit (EDU). The EDU allows the recording of an extensive number of different parameters during an inflight experiment. Not only the operational parameters, like position of the aircraft, altitude or speed are recordable, additional parameters e.g. concerning the use of the Flight Control Unit (FCU) and the strength of sidestick movement can be recorded in real time during the simulated "realistic" flight scenario.

Another important research capability is the ISCAN HEADHUNTER system, which allows the recording of the pilots field of vision at a rate of 50 Hz. Using this ISCAN system it is possible to monitor the direction of the pilots field of vision, the precise point at which he looks, the pupil diameter and the fixation duration of different points in the cockpit.

Method

In a first experimental approach, the workload and scanning behaviour including eye movement and fixation duration of pilots during flight in the simulator are studied. The aim of this study is to provide some empirical data about the scanning behaviour of pilots during flight as reference for further studies to evaluate the implementation of an experimental integrated ATC message display.

A total of five pilots were selected in this first series of tests. They all had commercial pilot licences and were first officers or captains on the Airbus A320. Their flying experience ranged from 1400 to 13000 hours on commercial aircraft. Unfortunately, no pilot with A340 experience was available. But the A320 and A340 have basically a common cockpit layout with similar flight operational procedures. Furthermore the experiments included only standard procedures without abnormal or emergency procedures making reference to the particular aircraft systems, therefore the selected test persons could be regarded as suitable in regard to the expected results.

In order to provide a suitable environment, a simulated flight task was selected. The task consisted of two 40-minute flights from Munich (Germany) to Salzburg (Austria) and back under normal operational and meteorological conditions. The two scenarios contained different levels of workload during take off, climb out, cruise, descent and ILS respectively NDB-DME approaches. For each scenario during half of the time the auto pilot system was disabled to require the pilot to fly manually using the sidestick. The flight director was usable.

Each test pilot had to perform the two described scenarios, first as pilot flying (PF) and on the return flight as pilot non flying (PNF) according to standard crew coordination concepts. The standard task sharing between PF and PNF means that PF primarily flies the aircraft whereas PNF is acting as system operator and perform the communication with ATC.

The ATC messages in this first attempt were transmitted conventionally via audio channel (VHF communication). The ATC simulation embraces deviations, e.g. direct routings, radar vectoring and an unforeseen out of service message concerning the ILS, from the operational flight plan which was handed out to the pilot before each flight.

During the flight experiments two cameras permit an in-flight recording, one camera takes views from the whole flight deck, whereas the other camera was oriented to watch the F/O-seat which was also the seat the test person was occupying. Additionally, all conversation in the cockpit and between ATC controller and the pilots were recorded.

Parameters recorded during the experiments were, e.g. position of aircraft, altitude, speeds (true airspeed and indicated airspeed), engine power (%N1), selected kind of display at the navigation display, all actions taken at the flight control unit (e.g. selected speed and altitude) and movement of the sidestick.

After each of the two flights the workload of the pilots was measured using the conventional NASA TLX subjective workload assessment techniques.

Scanning behaviour of pilots in modern glass cockpit

To analyse the scanning behaviour the flight was divided into 5 phases : Take Off (line up to 1500 ft GND), Climb (1500ft GND to FL160), Cruise, Descend (FL160 to 5000 ft) and Approach/Landing (5000 ft to touch down).

The analysis of the scanning behaviour was generated dependent on the actual flight time. For analysis of the viewing area, nine different locations were defined e.g. Primary Flight Display, Navigation Display, Flight Control Unit, etc. .

Overall, the PF's spent more than 65 % of the whole flight time scanning the Electronic Flight Instrument System (EFIS) (s.a. Waller et. al., 1989), whereas 35% were dedicated to the PFD and 30% to the ND. For further analysis fixation duration time, cycle time and the length of the cycles were used. The fixation duration time is the time the pilot monitors an object without leaving a viewing sector of 1 cm².

A cycle is defined by a number of different viewing areas the pilot scans without scanning the same area again. For example, if the pilot scans the PFD and afterwards the ND and then comes back to PFD, this scan pattern is defined as a three step cycle (PFD-ND-PFD). The cycle time is the time within the pilot completes one cycle. The length of a cycle is defined by the number of areas that a cycle contains, e.g. the length of the cycle mentioned above is three.

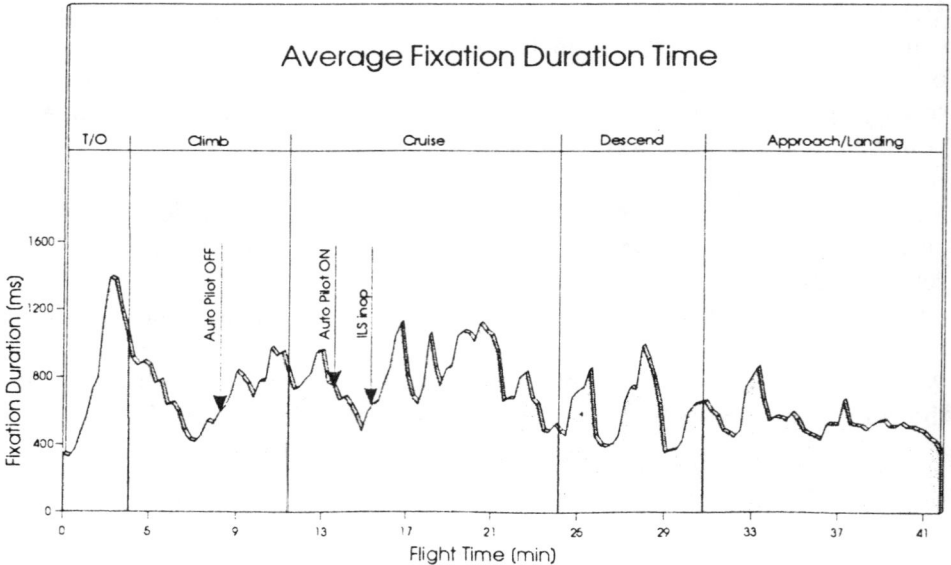

Figure 1   Average Fixation Duration Time

Figure 1 shows the fixation duration times as a function of the actual flight time. An increase of the duration time is found in the second half of the Take Off Phase (after lift off) and after the autopilot was inoperative. During the approach/landing phase a lower level of duration time was recorded.

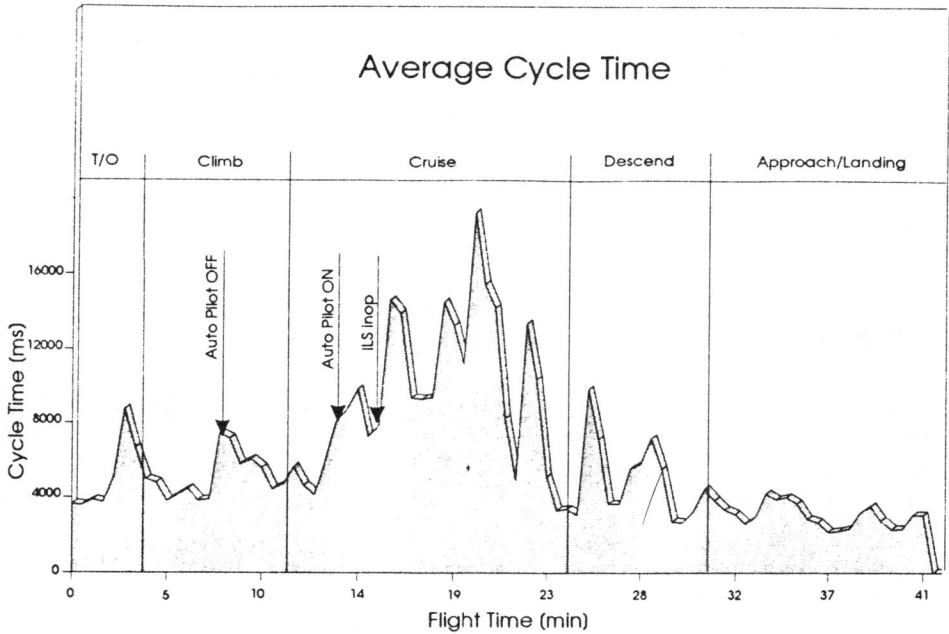

Figure 2   Average Cycle Time

Figure 2 shows the average cycle time as a function of the flight time. The cruise phase is characterised by a higher level of cycle time compared to the other phases. The time the pilot uses to end a control cycle, which means until he reaches the same viewing area (e.g. the PFD or the ND) again, is significantly higher in the cruise phase than in other phases of the flight. Especially the take off phase and the approach/landing phase is characterised by low cycle times. The average length of a cycle over the whole flight time is 3.6. This indicates that a typical scanning cycle includes only two different viewing areas before reaching a redundant area.

Design Consideration for an ATC-Display

The question whether using an existing display in a modern glass cockpit or implementing a new display for ATC messages should be answered as follows: Concerning the scanning behaviour, respectively the average length of a cycle, it is not feasible to implement a new display by creating an additional viewing area. The new display should be integrated in one of the existing EFIS displays because this area is in the main field of sight at least 65 % during the normal flight. The two EFIS displays differ especially in the way that the ND is presenting the more stable information, whereas the PFD presents information that underlie rapid changes. Besides, tactical ATC messages refer mainly to navigational flight guidance information displayed on the ND. In consequence, the ND seems to be the one instrument feasable to integrate an ATC message display in regard to the pilots human factor characteristics.

Prospect

Objective of these tests were initial data for the implementation of an ATC message display in a modern glass cockpit.

The project aims toward the development of an experimental integrated ATC message display, where the visual presentation of ATC messages is integrated into the existing flight displays in the cockpit. The data exchange will be realised via a simulated Mode S channel. In particular, aspects of the human machine interface are addressed, e.g. acceptance procedures, visual perception issues and implications on the pilot workload.

In a second step an experimental ATC message display will be integrated into the A340 cockpit and a simulated Datalink will be established using the A340 simulator. In this second study the scanning behaviour and in flight workload, using the new ATC message display, will be monitored.

Acknowledgements

The research of the above described study is funded by Deutsche Forschungsgemeinschaft (DFG), project reference Hu 345 / 4-1.

References

Bohr, T. (1990), 'ATM-Kooperativer Weg in die Zukunft', in Deutsche Gesellschaft für Ortung und Navigation, *Symposium Auswirkungen neuer Technologien auf die Sicherheit im Luftverkehr*, TÜV Rheinland.

Waller, M. C., Lohr, G. W. (1989), *A Piloted Simulation Study of Data Link ATC Message Exchange*, NASA TP 2859.

14 ERATO: cognitive engineering applied to ATC

M. Leroux, CENA, Toulouse, France

Introduction

The European ATC system has to face a tremendous increase of the demand. A large discussion on how to enhance ATC methods and tools is open. Very ambitious goals are assigned to the future systems.

Obviously, major technology improvements (FMS, Data Link, 4D–Navigation, computational power) must be intensively used. But, in the mean time, a full automation cannot be a solution, at least for the next two or three decades. Human controllers must remain in the decision making loop. As automation cannot replace human operators, it must assist them. As long as full automation feasibility and efficiency will not be proved, ie as long as we will need controllers to make decisions, even in an intermittent way, it is essential to preserve the controllers' skills. Whatever the tools that will be designed, human controllers must exercise their skills continuously.

Human operators are a factor of flexibility, of capability to deal with unexpected situations, of creativity, of safety, thanks to their capability to compensate for machine's failures or inadequacies. To preserve these capabilities, we may have to automate "less" than possible from a pure technological point of view.

But, in the mean time, the human operators are a factor of error. From this observation and for years, system designers though that the more human operators will be put on the fringe, the more the risk of error will decrease. In fact we add another kind of difficulty to the supervision of the initial system : the difficulty of understanding the behavior of the automatisms that partly monitor the system. Thus, automation makes the operators loose their skills, as they know less on the initial system. It creates additional sources of errors ; as reported in numerous exemples, the consequences of these errors are much more important than the previous ones. Better than eliminating human operators with the consequences of depriving the joint system of major benefits and of increasing the risk of errors, it seems more sensible to design a system which is error–tolerant. Such a system cannot be designed only from the technical advances : we must automate in a different way that suggested by technology alone.

Following is a short description of ERATO (En Route Air Traffic Organizer), a project from CENA. This project is aimed at designing decision aids for Air Traffic Controllers. It will result in a Controller's Electronic Assistant (CEA). The Electronic Assistant will be put in operations by the year 1999.

The general approach of ERATO

Central to this project is a cognitive engineering approach. Figure 1 represents the different steps of this project.

The cognitive model of the controller and the assessment of bottlenecks in decision making mechanisms

The first step in building a cognitive model is to determine what really makes the task difficult (Woods 1988). We have identified three sources of difficulty. The first ones are inherent in the physical process itself : controllers have to make decisions from data which depend on time for their value, their accuracy, their availability and their flow. Some other sources of difficulty are bound to cooperative activity. The third sources of difficulty come from the interface itself. In the mean time, the interface is a "window" on the real world and the only means to act on it. It is now recognized that an inadequate interface, or inadequate tools, may be a critical source of complexity.

The cognitive model explicits the mental mechanisms and resources which enable the controller to face this complexity, and describes how these mecha-

90

nisms decay under time pressure, stress and fatigue (Leroux 1993). Following is a brief description of one point of this model.

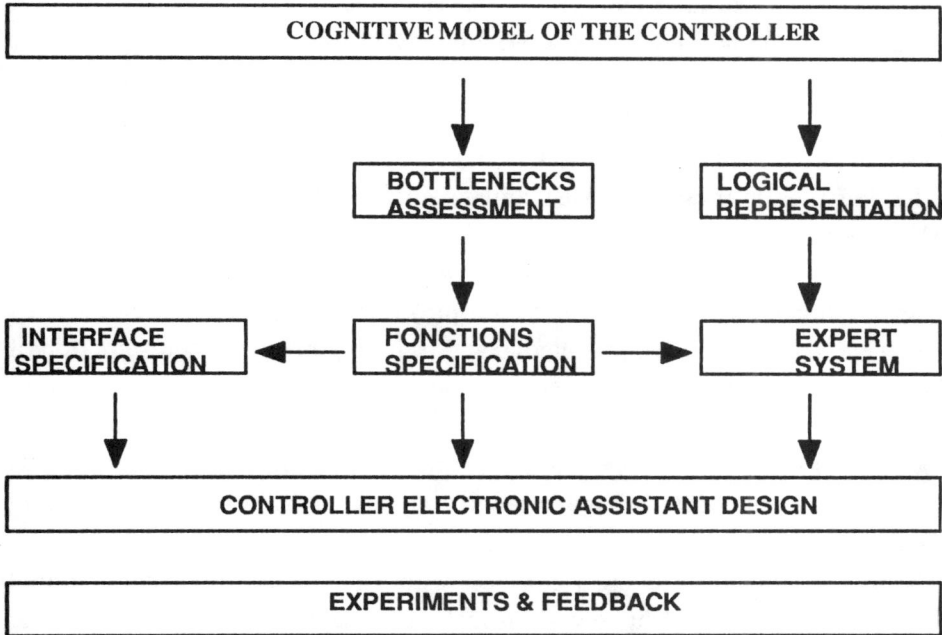

Figure 1 Global architecture of ERATO

When the controller does not know relevant data to make decision, for example a top of descent, he anticipates according to a "normal", routine, behavior of the aircraft called the default world, with reference to the "default logic" that models this kind of reasoning. Default reasoning is a means of making decision in a state of partial ignorance. This is an efficient means of narrowing the range of contingencies to be examined and to increase efficiency : at first all the aircraft are processed as if their behavior always remain consonant with the "normal" behavior.

But to process exceptions, that is to guarantee safety, controllers monitor "sentry parameters". As long as these parameters remain in a "normal" range, all the previous diagnosis or decisions that are inferred from the default world, remain valid. But if a sentry parameter drifts outsides the expected range, than all the previous plausible inferences have to be revised : some additional conflicts can be created, due to this abnormal behavior. But we can observe that, in very demanding situations, the monitoring task may no longer be performed by the controllers. Thus, when outside its validity domain, ie in too demanding situations, this mechanism may become a major source of errors.

In the same way the cognitive model explicits mechanisms that enable the controllers to eliminate ambiguity, to trigger relevant resolution frames from the assessment of the problems, to choose amongst these frames, to instanciate the choosen frame, to abandon a frame and to shift from one frame to another one. The model also addresses the memorization problems, due to frequent shifts from one problem to another one, as well as to frequent and unanticipated interuptions of mental processes. Mechanisms enabling to trigger attention at the right time to perform the right action are also very costly.

Mental mechanisms involved in cooperation are an essential part of the model. Efficient cooperation between the two controllers relies on three factors. They must have :the same skills, knowledge and training, the same representation of effective traffic requirements, and simultaneously available cognitive resources to exchange information. When demand increases, these two latter conditions may decay so much that cooperation may no longer be effective. Numerous airmisses have been reported, that are due to cooperation failure in too demanding situations.

All these mechanisms severely compete with. As the controllers have limited cognitive resources, they have various techniques and mechanisms that enable them to manage their own cognitive resources. The description of these mechanisms is central in the model.

Improving the efficiency of all these mental mechanisms suggests the use of problem driven information filtering

The information filtering module of the Electronic Assistant

When considering an aircraft, the minimum set of relevant data for the controller to manage with this flight is composed of the subset of all conflicting aircraft and the subset of all the aircraft that may interfere with a radar resolution of these conflicts, that is all aircraft that may constrain conflict resolution.

The filtering module processes the same set of data as the controllers have to process now, ie the information from the strips and, when available, the radar information. It includes two main modules. The first one computes the default representation of each aircraft. From this representation, the second module associates to each aircraft its relevant environment, called interfering aircraft subset (IAS). This subset is not determined by means of a pure mathematical computation, but according to current expertise of controllers.

Information filtering techniques are under dispute (De Keyser 1988). The point is how to make sure that the operator won't need a data that is hidden by the system ? Such a data retention should be an unacceptable source of errors. The discussion on the exhaustiveness and the relevance of data filtered by the filtering module is central.

The first answer consists in taking into account the default behavior of the aircraft in a more "prudent" way than the controller. It will result in the display of some aircraft that may be not relevant for the controller. But the aim of information filtering is not to provide the controller with the exact subset of relevant data, but with a subset of data including all relevant datais and easily manageable.

But this first answer does not really solve the problem. The knowledge elicited in the filtering module defines a set of "normal behaviors" of the controllers. But it is impossible to represent the whole knowledge of all the controllers. Should we be able to do this, that we should have to deal with controllers'errors or creativity. The solution defined in Erato consists in considering the filtering module as a default representation of the controllers. To guard against the consequences of human error or creativity, (ie unexpected behavior) a monitoring process is associated ; this process is inspired by the natural sentry–parameters monitoring process of the controllers. This monitoring process will detect all discrepancy between the actual position of all aircraft and any of the possible position as it could result from a "normal" behavior of the controller. When necessary this process will trigger an alarm, so as to advise the controller that the previous information filtering is no longer relevant and has been updated.

This monitoring process associated to the filtering module allows the electronic assistant to adapt very smoothly to operator's error and creativity. Such an information filtering is error tolerant.

How the controller's electronic assistant uses filtered data

The problem–driven information filtering allows the controller to focus all his activity on well–formulated problems ; so he can operate all his mental mechanisms in a more efficient and creative way. This function substitutes a set of easily manageable problems to the initial complex situation.

The basic information filtering will be used by some additional functions such as the extrapolation function, the simulation functions, some memorization aids, and data–transfer functions which are aimed at improving cooperation between controllers on the same control position.

The reminder consists in a specific window of the Electronic Assistant, where each problem will be associated a label. A problem is defined as a conflicting situation involving two or more aircraft. The labels are positioned according to the urgency. The display of the relative urgency of problems should enable the controller to avoid wasting cognitive resources on non urgent and non important tasks while the short term situation decays. In normal operation, this should allow him to manage more objectively his own cognitive resources.

Validation aspects

The validation process is widely described in Leroux (1992). The validation of the filtering module is already in progress. It will involve 40 controllers. Large scale experiments of the Electronic Assistant and of the joint Man–Machine system will involve 20 controllers for 5 monthes by the end of the year 1994. These experiments are the initial phase of an iterative process leading to final specification of the Electronic Assistant by mid–1997. The cognitive model will be a guideline all along these experiments and , in return, it will be improved.

References

De Keyser, V., (1988). – L'ergonomie des processus continus. De la contingence à la complexité: l'évolution des idées dans l'étude des processus continus, *Le travail Humain,* 51, 1–18.

Leroux, M. (1993), *The role of expert systems in future cooperative tools for air traffic controllers,* 7th International Symposium on Aviation Psychology (Colombus Ohio 1993).

Leroux, M., (1992).– The Role of Verification and Validation in the Design Process of Knowledge Based Components of Air Traffic Control Systems in Wise,J.A., Hopkin, V.D.& Stager, P., eds. *Verification and Validation of Complex and Integrated Human–Machine Systems.* Vimeiro, Portugal : NATO Advanced Study Institute proceedings.

Woods,D.D.,(1988). – Commentary: Cognitive engineering in complex dynamic worlds. *in* Hollnagel,Mancini,Woods (Eds.) *Cognitive Engineering in Complex Dynamic Worlds.* London, Academic Press, Computers and people series.

15 A stress-based analysis in Air Traffic Control

A. Bellorini and F. Decortis
Institute for Systems Engineering and Informatics, Joint Research
Centre, Commission of the European Communities, Ispra, Italy

1. Introduction

A field study in Air Traffic Control (ATC) has been conducted at the Milan airport where activities of Air Traffic Control Operators (ATCOs) were examined during the phase of "Approach Control". The research aimed at the analysis of causes and effects of stress on ATCO cooperative work, distributed cognition, attention and communication. We want to investigate and describe operators' tasks and to relate the results obtained from these analyses to improve the conceptual basis for designing better information support tools and analysing the impact of information technology on work conditions. The decision to carry out a research in the ATC domain is motivated by a 50% increase of traffic during the last five years. It is therefore important to understand what effect this increase might have on task performance and cognition. It appears that in stressfull conditions the operators need more adapted support tools to accomplish their tasks. Many critical aspects of the task, as well as weaknesses in the design of supporting tools, manifest themselves under stressfull conditions.

Previous studies in the ATC domain have revealed that the most important causes of stress are the workload (Crump, 1979) and the temporal stress (Moray, 1982). Other studies suggest a relationship between workload, number of aircraft controlled and number or duration of communications (Hopkin, 1971). The ATC task appears to be a cooperative activity. Much of the activity consists in the ability to organise the distribution of individual tasks among the team into an assemblage of activities within the working division of labour (Hughes & al. 1993; Bentley & al., 1992) and this also in order to cope with stress. Recent developments in work psychology, social science, namely Ethnomethodology and Conversation Analysis, provide a theoretical and methodological framework to explore the social and cognitive aspects of collaborative work. The communication is both a means and an indicator of the way in which operators work as a team and of the collective managment of the cognitive resources. The main concept implied in communication is that agents overtly reach a situation of shared mental states and beliefs (Airenti et al. 1993). In the ATC task we have a conversational cooperation in terms of establishment and control of sharedness throughout conversation (Clark& Schaefer, 1989; Clark and Wilkes-Gibbs, 1986).

2. Experimental methodology

In this research we have applied the Ethnomethodological approach related to the analysis of the every-day activities and, in particular, the Exploratory Sequential Data Analysis (ESDA, Sanderson, 1993). ESDA includes the analysis of the recorded data in which temporal information has been preserved.

Eleven Radar Controllers from "South Approach Control" have participated in the experimental research. A Cognitive Task Analysis obtained through interviews and observations has been carried out to collect enough knowledge to understand ATC task and to develop a methodology for the collection of data. The cooperative work involve Radar Controllers from several sectors as well as Pilots. The ATCOs must dynamically process large amount of information from various sources to decide on an action within short time periods. Different cognitive processes are involved in ATCO task such as memory, attention, distributed decision making, anticipation and cooperative work based on mutual knowledge.

2.1. Data collection

The data collection consisted of the following elements:
- *observations*, during the video-recording, of the situation controlled by the South Radar Controllers such as the number of aircraft in holding, landing and taking off, the weather conditions, the parking available in the airports and some unexpected problems;
- *video-recording* the South Radar Controllers during the execution of their tasks, under high traffic conditions, in order to derive some mechanisms related to the "face-to-face communications" between North and South Radar Controllers, radar interaction, strip manipulations and environmental conditions (noise, lights, equipment);
- *audio-recording* the communications between South Radar Controllers and Pilots and between South Radar Controllers, Adjacent Sectors and Tower Controllers in order to obtain verbal protocols;

96

- *self confrontation* with video-recording including South Radar Controllers verbalizations and explications on their cognitive activities and strategies;
- *subjective rating scale of stress* during the self confrontation at appropriate temporal intervals concerning the South Radar Controllers subjective measure of stress according to their experience and stress tolerance.

2.2. Data transcription

McSHAPA has been used for analysing verbal protocols (Sanderson, 1992). This tool provides a standardized framework to represent verbal and non-verbal behaviour that is amenable to a variety of data reduction and analysis techniques. We have transcribed within McSHAPA (fig.1) the observations concerning the context and the time of each situation recorded; the "distance communications" between Pilots and South Radar Controllers; the "face-to-face communications" between North and South Radar Controllers, Tower Controllers and Adjacent Sectors Controllers; the Controllers' explanations and comments about their strategies and difficulties; the rating scale of stress of the Controllers.

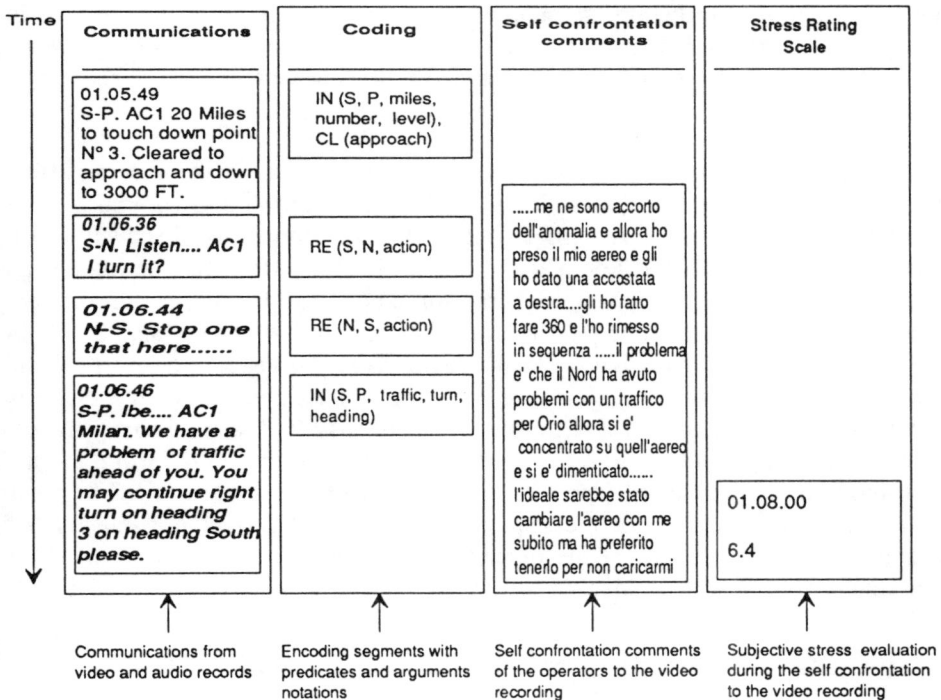

Figure 1. Extract from "column" representation of using McSHAPA.

2.3. Data analysis

A stress-based analysis has been applied in order to reduce the large amount of data. The objective of the analysis is to identify causes of stress and to analyse their consequences on cooperative work, in particular the distributed decision making process, attention and communication. The elements of a stress-based analysis are (1) video-based analysis, to identify stress related events; (2) quantitative analysis of measures of stress on a subjective rating scale; (3) "narrative analysis" along the stressful events; (4) stylistic and content analysis of the phraseology, to identify sensible index of stress; (5) Verbal Protocol Analysis (Ericsson and Simon, 1984).

3. Results

The main causes of stress appear to be related to the *number, the kind and the time* of communications from different agents which can overlap on the same source (headphone) or on different sources (headphone and external voice). The *temporal distribution of the task* is another cause of stress as well as the issue of traffic that is controlled at the same time in different zones of the area (temporal overlapped tasks). The causes of stress related to *traffic characteristics* are the complexity of the order with which the aircrafts enter the sector, the volume of the aircrafts in charge and the holding aircrafts for landing on the different gates. The appearence of *unexpected problems* represents unplanned actions to be taken under short time constrains.

The case of conflict resolution where temporal overlapped tasks have an effect on cooperative work, distributed decision making, attention and communication merit further attention. This case is related to the cooperation beetwen North Radar Controller (NC) and South Radar Controller (SC) (fig. 2). The SC has an aircraft, AC1, leaving a gate (Voghera) for landing at the Milan airport. SC decides the order of landing of the aircrafts coming fom North and South in cooperation with NC. SC gives AC1 the flight level, the heading orientation, the number of landing and the clearance to approach. At the same time NC has different aircrafts to control and, in particular, is busy with a traffic from Nord-East landing in another airport. He controls an aircraft, AC2, from a gate (Saronno) landing at Milan airport. He has to transfer this aircraft to SC in order to take a shared decision for the sequence of landing. NC does not transfer the aircraft anticipatly because of the other traffic and lets the AC2 go ahead. As a consequence SC has to suddenly redirect AC1 in order to avoid a conflict with AC2.

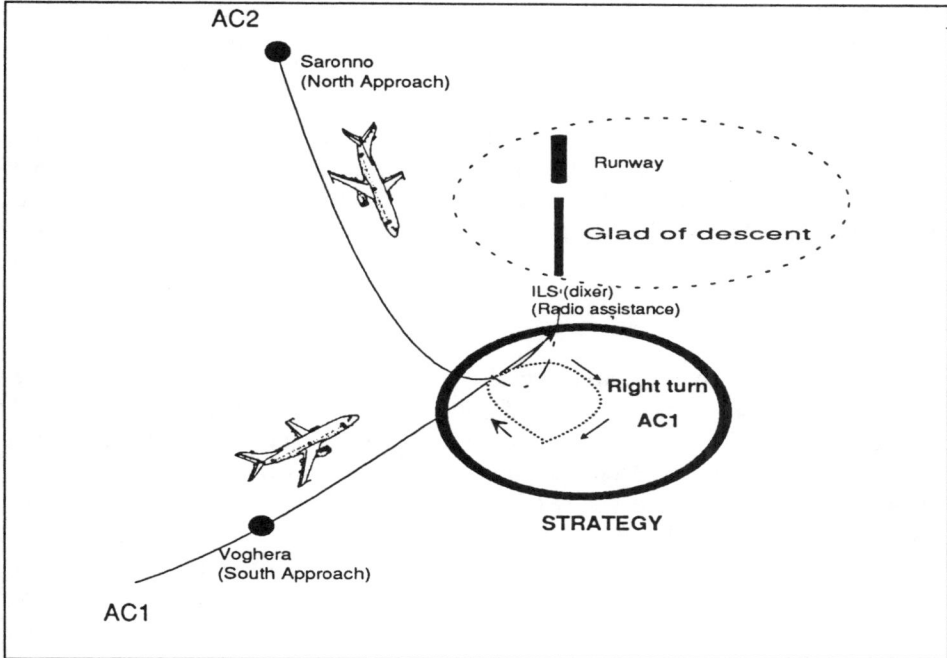

Fig. 2: Representation of the situation controlled by North and South Radar Controllers: a conflict resolution.

We observe here the shift from cooperative work to individual work, and the impossibility to carry out a distributed decision making process. The problem ends with the action of SC on his aircraft (AC1). We can explain this alteration of cooperative work in this way:

- NC is busy on another traffic and the high number of communications related to this traffic take all his attention. The resources are allocated on the problem (problem driven) acccording to the weight of the workload and the temporal pressure.
- NC cannot reach the distributed decision making to pass over anticipatively his aircraft to SC because of the lack of time and resources; he tried to wait until this can be possible to act in cooperation.
- SC knows the workload of NC and does not ask to control AC2; the cooperative work is realised through the "face-to-face communication" that increases the workload. The distribution of the task is based on the mutual knowledge of the workload.

Under stress, the attention is focused on the problem (problem driven): the Radar Controllers's distribution of attention on the different sources of information which operates in normal situation is altered. The radar screen and the "distance

99

communications" became the main sources of information. Memory is overloaded. The Radar Controllers uses the flight strip as memory support for the flight level of the aircraft in holding. We have observed also a temporal restriction on the anticipation. The analysis of the communications have resulted in the definition of some objective index of stress: errors such as repetition, partial and incorrect readback and misunderstanding. The syntactical rules are not respected and there is an increase of the mutual and tacit knowledge: nick names and abbreviations are frequently used in the communication.

4. Conclusion

Results show that the subjective measure of stress can be related to objective parameters such as the number and kind of communications, the temporal distribution of the task, the traffic characteristics and the appearence of unexpected problems. These elements have been shown to impact on the performance of cooperative work, the distributed cognition, the selection of the focus of attention, the communication, the memory load and the anticipation. These results suggest that the current control tools used for the managment of stressfull situations can be improved. Also, the critical aspects of the ATCO tasks, including the needs of the ATCOs to carry out these tasks should be taken into account during the design of future control tools.

References

Airenti, G. Bara, B. G. Colombetti, M. (1993). Conversation and behavior games in the pragmatics of dialogue. *Cognitive Science*, 17, 197-256.

Bentley, R. Hughes, J. Randall, D. Rodden, T. Sawyer, P. Shapiro, D. Sommerville, I. (1992). Ethnographically-informed systems design for air traffic control. *Proceedings of Conference on Computer-Supported Cooperative Work*, Toronto, Canada.

Clark, H.H. & Shaefer, E.F. (1989). Contributing to discourse. *Cognitive Science*, 13, 259-294.

Clark, H.H. & Wilkes-Gibbs, D. (1986). Reffering as a collaborative process. *Cognition*, 22, 1-39.

Crump, J. (1979). Review of stress in air traffic control: its measurement and effects. *Aviation, Space and Environmental Medicine*.

Ericsson, K. A. Simon, H. A. (1984). *Protocol analysis. Verbal reports as data.* The MIT Press, Cambridge, Massachusetts.

Garfinkel, H. (1967). *Studies in Ethnomethodology.* Prentice-Hall, Inc., Englewood Cliffs, New Jersey.

Hopkin, V. (1971). Conflicting criteria in evaluating air traffic control systems. *Ergonomics*, vol. 14, n.5, 557-564.

Hughes, J. Randall, D. Shapiro, D. (1993). From ethnographic record to system design. *Computer Supported Cooperative Work*, 1, 123-141.

Lazarus, R. (1979). Psychological stress in the workplace. *Journal of social behaviour and personality*. Vol.6, n.7.

Sanderson, P. M. (1993). McSHAPA and the enterprise of Exploratory Sequential Data Analysis (ESDA). *Manuscript submitted for publication to International Journal of Man-Machine Studies*.

Sanderson, P. M. Fisher, C. (1992). Exploratory Sequential Data Analysis: traditions, techniques and tools. *CHI'92 Pre-Workshop Report*, Monterey, Ca. Engineering Psychology Research Laboratory, Department of Mechanical and Industrial Engineering, University of Illinois at Urbana-Champaign.

16 The human role in aircraft–air traffic automation integration: what we don't know can't help us

Russell A. Benel, Center for Advanced Aviation System Development, The MITRE Corporation

Recent system development activities have exposed the essential need for information to support meaningful system integration decisions resulting in systems that employ effectively both flight deck and air traffic control (ATC) personnel. Researchers and developers often limit analysis, design, and evaluation efforts to the human role within an individual system domain (often a single operator), not the integrated system encompassing relevant domains. International developments further imply the need to create a "seamless" system where global aircraft operation is enhanced by the automation encountered and each system's personnel are supported effectively despite differing characteristics of the systems installed at these locations and in different aircraft. In this paper I will identify (and provide examples of) the early use of system simulation prototyping to derive human performance characteristics applicable to the design and concept validation of both individual and evolving, integrated air traffic management systems.

Background

Earlier flight deck and ATC automation programs had elements that were relatively well-understood and specified, designed, built, tested, and fielded independently in a straightforward manner. Now, the opportunity to develop systems with greater levels of capability introduces both increased complexity and interaction among system elements. There is a much wider diversity of aircraft and equipage, more sophisticated cockpit displays and flight management tools, and increased automation of airline schedules. Delayed delivery of new capability to the field, observed unintended consequences of automation and its impact on operational personnel, and cost growth for these recent upgrades clearly demonstrate that system development methods of the past are not successful in today's more complex environment.

The international aviation community needs an approach to building airborne and ground-based automation systems that interact to support operational personnel and deliver safer, more efficient ATC system services. System level prototyping allows study of components in the context of the entire system anticipated to exist when each component is implemented, and supports:

- Concept validation at the system level, focused on user needs;
- Future systems engineering, planning and development, studying how system elements (including functions allocated to humans) will work together and how they will deal with unplanned events; and
- Human factors assessments in a highly integrated future automation environment.

In addition, successful system prototyping provides direct input to the system development process ensuring that the right systems are built and they are built right. This prototyping involves all relevant disciplines as integral participants (see Figure 1).

System issues and concepts

In general, systems are an assemblage of parts forming a coherent whole with an objective. The major system elements of the ATC system (both aircraft and ground-based ATC) comprise many other systems. Aircraft are evolving to include system elements that provide complementary and, in some cases, redundant functions with those on the ground. Flight Management Systems, Airborne Weather Radar Systems, and Traffic Alert and Collision Avoidance (TCAS) on the aircraft have analogous functions in the ground-based systems such as AERA (formerly Advanced En Route ATC) and the descent advisor

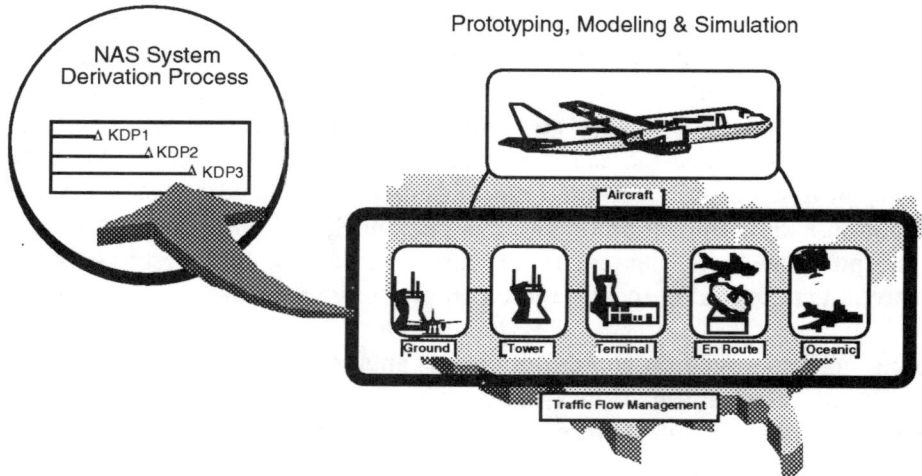

Figure 1   System simulation results influence system development decisions

(DA) function of the Center TRACON Advisory System (CTAS), Weather and Radar Processor (WARP), and conflict prediction and detection functions in a variety of the ATC systems in the United States. It is the integration of this emerging set of systems capabilities that represents the basic challenge and opportunity for the definition of an effective role for the human operators.

ATC and flight deck systems exhibit nearly continual change throughout their life span. In addition, both the number and type of momentary elements of the system are not fixed. The particular aircraft flying, ATC facilities (and positions) in operation, and operational environment (e.g., weather and traffic) may all change over a short time. These factors lead to continual adjustments on the part of the ATC participants (i.e., airlines, pilots, ATC, and other operational personnel). Thus, ATC system elements, airspace structure, aircraft, procedural, infrastructure, and behavioral evolution are inherent characteristics. Behavioral evolution provides a significant challenge to the definition of an effective human role. In many ways, this natural adaptation is out of control of the system designer and is difficult to anticipate. A significant challenge is to acknowledge and harness the adaptive power of humans within the integrated Flight Deck-ATC system.

What we don't know

Two classes of information seem difficult to derive from standard human factors design sources, both may be addressed through simulation. The first is technology driven, characterized by emerging user interface techniques, and related functions. The second is inherent to system evolution. Namely, what is the expected long-term performance of personnel, shifts in roles over time, and system performance under the new "steady-state" implementation.

In the past, automation was designed specifically for each application. Today, there is increasing pressure to apply commercial-off-the-shelf (COTS) products (for cost avoidance) and incorporate their user interface techniques (representing modern computer human interface, CHI, look and feel). Thus, our tasks include reconciliation of (slightly) incompatible CHIs and associated modes of interaction and evaluation of the efficiency of commercial CHIs applied to real-time systems. Interaction with information databases and electronic presentation of manuals rather than printed data will raise issues of storage, retrieval, and organization for optimal presentation. Application of alternative input/output will require additional design effort for these application, e.g., error editing for voice entry is used or speech output quality.

New interfaces to perform identical tasks alone should cause evolution of human roles. Moreover, shifts from manual tasks (e.g., electronic flight data replacing printed strips) have been called into question (see Hopkin, 1992). Although we have no evidence that previous methods supported mental model and situation awareness formation, we know that the inevitable change can be disruptive (temporarily). Changed functions ensure that tasks will change for individuals, teams within a domain, and the interaction among individuals and teams in the various domains, but we don't have accurate, current information as a baseline of comparison for most performance.

One recent example is the subtle shift toward use of the traffic information on the TCAS display to enable other actions such as early execution of fuel saving maneuvers, spacing to reduce traffic congestion, and enhancing visual approaches (see Mundra, Cieplak, Domino, and Peppard, 1993). Each addition and change to the Aircraft-ATC automation system bears similar potential. We must anticipate change even when we can't predict it exactly.

The role of system simulation

The United States Federal Aviation Administration (FAA) established the Integration and Interaction Laboratory (I-Lab) as part of the National

Simulation Capability (NSC) to support systems throughout the development process for the future air traffic system. Our studies have helped define future system requirements, reduced system interoperability risk, supported end-user evaluations of emerging functionality, and expedited the delivery to the field. Early use of system simulation prototyping is essential to derive human performance characteristics applicable to the design and concept validation of both individual and evolving, integrated air traffic management systems. Benel and Benel (1994) noted that you always build a prototype, but, sometimes, everyone is under the illusion that it is the real system. Effective application of results should limit the chance that the system becomes a mere prototype.

Morgan, et al. (1992) outlined the FAA's framework for systems engineering with renewed emphasis on user needs and systems evolution an iterative and interactive process including system users, operators, and maintainers. Although we may still see concomitant change in the operational environment as new technologies are fielded, system simulation can eliminate the need for expensive, time-consuming changes common during full-scale development. Envisioning how the emerging system might be used through system simulation anticipates changes in procedures and related operational concepts, additional system changes to enable full benefits, and timing of delivery of integrated system elements. Thus, the prototype can become the system.

Studies and exercises

En route-Terminal-Traffic Flow Management automation integration studies have allowed us to examine parameters and operational concepts developed for application within each domain. This has allowed redefinition of parameters and implementation concepts for effectively integrated systems to support arrival traffic management. Integration of airborne automation capability and data link services will complete the system simulation capability.

Exercises to illustrate the potential benefits of alternative procedures have expedited change in Oceanic airspace (see Mundra et al., 1993) and alternative operational concepts for integrated systems. Initial exercises lead to detailed flight simulator trials and full mission ATC simulations prior to flight demonstrations resulting in a new oceanic procedure for TCAS assisted In-Trail-Climb in less than one year from concept to fielding.

Evaluation of a planned addition to the current system has included design of data link application for downlink of a TCAS resolution advisory message to the terminal controller. This would allow the controller to know why an aircraft has deviated. Issues include the timing and format of messages and utility of the application.

Use of the data

Jenny and Fearnsides (1993) have proposed that it is possible to implement a prototyping approach within the acquisition process without any change in the US Federal Acquisition Regulation, but program managers and acquisition officials must be willing to accept appropriate alternatives within the regulations. In some cases, prototypes may become the system. More commonly, the data from the prototyping exercises may improve procedures, aid in the decision making process, or alter a programs characteristics or timing for system benefit. Operational personnel are enabled by prototyping. It is possible to derive greater insight and provide better feedback on potential system designs. In the ATC system context, this may include pilots, controllers and related operational personnel sharing insights.

Acknowledgment

The opinions expressed are solely the author's and do not represent those of The MITRE Corp., previous employers, or current or previous sponsors.

References

Benel, R. and Benel, D. (1994), A Systems View of Air Traffic Control, in M. Smolensky and E. Stein (eds) *Human Factors in Air Traffic Control.*, New York: Academic.

Federal Aviation Administration. (1990), *The National Plan for Aviation Human Factors.* Washington, DC: Author.

Hopkin, V. (1992), Human Factors Issues in Air Traffic Control, *Human Factors Society Bulletin*, 35(6), 1-4.

Jenny, M. and Fearnsides, J. (1993), Management Strategies for a Changing Air Traffic Management System. *38th Annual Air Traffic Control Association Conference Proceedings*, 22-A.

Morgan, R., Lowenstein, J., ElSawy, A., Sinha, A., and Willingham, F. (1992), *FAA Systems Engineering Management: Renewed Emphasis on User Needs and System Evolution* (MP 92W0000037). McLean, VA: MITRE

Mundra, A.D., Cieplak, J.J., Domino, D. and Peppard, K (1993), Enhancing Visual Approaches with the TCAS Traffic Display. *38th Annual Air Traffic Control Association Conference Proceedings*, 582-588.

17 Human Factors issues of advanced ATC systems

Lambros Laios and Maria Giannacourou
Department of Industrial Management, University of Piraeus,
Piraeus, Greece

Introduction

Air traffic control is a complex task with the main objective to ensure the safe, orderly and expeditious flow of air traffic. To achieve this objective numerous functions employing diverse facilities and aids are fulfilled collaboratively by air traffic controllers.

Systems currently in use have evolved over many years from basically procedural forms of control (which were suited to low traffic density, low workload environment) to the modern high traffic density, high workload tactical radar control experienced near major airports and airway junctions.

For several years automation in ATC has been advancing in the technological update of systems. During these years exposure with automated functions has provoked an increasing emphasis on the "human factor". Human factors are concerned so much with human-machine interface aspects as well as with broader aspects such as the impact of automation on the role of the air traffic controller.

The present study is concerned with the possible implications of automation in ATC on critical human factors issues and especially the expected impact of specific ATC automated functions on tasks and role of controllers.

ATC automated functions

At present, a large variety of automated functions have been implemented or are under evaluation in advanced prototypes. For the purposes of the present work the following categories of automated functions were taken into consideration :

Electronic Data Displays : They provide electronic presentation of data under the appropriate form i.e. existing strip format or enriched format.

Trajectory Prediction Aids : These are 4-D predictions of the onward path of aircrafts based on the best information available to indicate deviations from the flight plan route and level data. It is envisaged that these function will support the acknowledgement of the deviation and the assessment of its criticality.

Conflict Detection Aids : They provide conflict alert using flight plan data enriched through radar updates and tracking predictions.

Data Links Applications : They include data link communications of weather data, air traffic control instructions and clearances as well as the automatic transition of data from on board navigation systems including aircraft identification and 3-D positions.

Clearance Advisory Aids : Such applications, through presentation of relevant flight data and potential conflict situations, aim to assist in the selection of the most appropriate entry and exit flight levels.

To understand the consequences of new automated ATC systems, it is necessary to understand the precise nature of the technological change occurring as a result of the implementation of such systems. Changes may affect human performance in different ways and although in the short term the improvements may be visible, other desirable features of human performance may be deteriorated.

Aspects of Human Performance affected by ATC Automation

Supervisory Control So far research in supervisory control (Barlini, 1981, Parsons, 1985, Sheridan, 1980) suggest that a possible outcome of the introduction of automated ATC systems is the reduction of routine activities. On the other hand, certain jobs are expected to be downgraded so that even greater increases in routinity will be experienced. Hopkin (1982) states that automation in ATC may increase routine actions if predictions, problem solving and decision making is taken over by the computer.

Monitoring Air traffic control is a complex monitoring task which is characterized by several dynamic displays each involving high information content and signals.

When tasks are automated, the controllers role becomes one of a monitor and a supervisor. It is this characteristic wherein the computer is assigned to receive information about the state of some ongoingphysical process and, based upon such sensed information as well as information programmed into it by a human supervisor, direct actions on that process, (Sheridan, 1980). Hence, the primary issues revolve around human ability to supervise this process since the task is almost always accomplished satisfactorily by the automatic system. Typical questions to be examined are whether automation breeds inactivity of complacency (NSTB 1976, Wiener, 1980).

Coordination In spite of the participatory nature of ATC task, the tendency in automation systems is for each man to become autonomous. Such an approach has numerous implications for collaborative effort. It has been noticed in evaluations of designs for future systems that team work sometimes breaks down as the system becomes progressively more loaded, so that tasks which are done collaboratively by a team under light workload tend to become fragmented into individual tasks under heavy loading, and each individual team member becomes too busy with his own task to keep his knowledge of his colleagues' activities up-to-date, (Kleinman, 1990). Thus, the team no longer functions satisfactorily as such.

Standardization Standardization of work practices is expected to increase with the introduction of automated ATC systems and is expected to apply as much to general work practices, eg. sectorisation of airspace, as to the homogeneity of used data. Standardization can have negative aspects. These may include loss of flexibility e.g. ability to respond in new unexpected situations and loss of job variety (Majchrzak et al. 1985).

Communication Another area highly affected by automation is communication. It can be hypothesized that automation can overcome traditional difficulties in communication allowing diversity and complementarity of communication channels, e.g. datalinks coupled with voice facilities. An estimated consequence of automated assistance in ATC is that since most aids are more suitable for individuals than for teams dialogues between human and machine will take the place of interactions between people.

Workload In the literature there have been identified two approaches concerning handling of controller's workload in order to increase ATC system capacity: reduction of workload through reduction of time spent on each aircraft and smoothing of workload.

A further suggestion to increase the controller's capacity is to smooth controller's workload. Two main methods have been identified for smoothing workload: a.) computer takes an active role in scheduling tasks, and b.) controllers are given the means to smooth their workload through greater control over the tasks, their timescales, their scheduling and their sequencing. Such approaches place the emphasis from tactical to strategic air traffic control and from the solution to problems to their prevention.

Expected Effects of Specific Functions on Human ATC Performance

The ATC functions can be divided into two groups according to their expected effects on human ATC performance : Trajectory Prediction, conflict Detection and Clearance Advisory aids form one group, whereas Electronic Data Display and Data Link Functions form a second group. The first group is envisaged to support tasks requiring : prediction, choice, and diagnosis, while the second group is envisaged to support tasks requiring : attention to alphanumeric information, acquisition and integration of new information, and decision on a response such as to change records on traffic. The specific effects of each function may be :

1. The use of Electronic Data Displays is expected to increase control actions. The replacement of flight strips with electronic displays and the consecutive substitution of writing on the strips with updating such displays is thought of requiring more input activities such as the use of keyboard, mouse or touch input devices to fulfil the same function. This increase can also be associated with the expected standardization of data and work practices in acquiring information and updating electronic data displays.

2. The use of Trajectory Prediction aids is expected to help controllers assess traffic situation better. Computer assistance in predicting airplane trajectories can enhance controllers' situational awareness through supporting comprehension of current situation and projection of future status enabling them to detect traffic patterns so as to determine which airways will be free and where potential collisions can be expected. But such a function can lead to inactivity due to complacency.

3. The Conflict Detection function can lead to increase in the traffic planning horizon. Memory requirements, mental effort and judgement errors are expected to decrease. A major positive consequence, here, can be the smoothing of air traffic (especially in medium term conflict detection) due to early problem diagnosis while among the negative consequences of using such aids can be a deterioration of controller detection skills due to dependence on computer function.

4. Data Link functions can be expected to reduce communication errors. Current research on data links reports fewer failures in the information transfer process (Kerns, 1990) such as assurance of correct interpretation of clearance or assurance of correct recipient (EUROCONTROL, 1991).

5. The use of Clearance Advisory may lead to the deskilling of controllers. It can be hypothesized that through the implementation of this function the ability of, especially, Planner controllers to develop problem-solving strategies and use knowledge about resources may subside since ready solutions will be proposed by the computer, and controllers job may change to simply accepting or rejecting proposed solutions. Also, in relation to the above consequence the use of such functions is expected to decrease mental effort, memory requirements and judgement errors.

In general, the introduction of automated functions can lead to an increase in formalization of data used and work practices as experience from relevant scientific fields has shown. It can be expected that due to automation the format of exchanged messages, sequence, pace and phraseology will become more standardized and systematization of procedures will be introduced eg. the use of formal procedures to complete work. However it still remains unclear whether the increase in formalization will decrease task flexibility or whether it will help in eliminating repetitive actions increasing thus speed of task execution and allowing time saved to be exploited in more critical tasks.

Automation of ATC tasks involving diagnosis and prediction (ie. trajectory prediction, conflict detection and clearance advisory) is also expected to alleviate pressure related with tactical planning, increase traffic planning horizon and optimise exploitation of available resources (airways, airports and personnel time).

The possible negative consequence of the automation in ATC associated with the functions of the first group can be the increasing dependence of controllers to automated functions resulting to loosing of control. Another negative consequence can be a significant increase in routine actions associated with updating traffic data relating to the functions of the second group.

In conclusion, the aim of this paper was to isolate and relate critical human factors issues with specific advanced ATC functions. The effects of automation on tasks and roles is not unilateral but depends on the nature of the specific function. However, in studying advanced ATC systems the net effect of automated functions on human factors issues integrated in particular systems should be investigated in longitudinal studies before final conclusions can be reached.

Notes

1. This research has been conducted in the context of the SWIFT (EURET 1.5) project funded by the CEC.

References

Barlini, P. (1981), The SCM case : The Experience Undergone by a middle Sized Mechanical Industry in the Field of Design and Automation Drawing Systems. In J. Mermet (Ed.), CAD in Medium Sized and Small Industries. Amsterdam: North-Holland.

Common Operational Performance Specifications (COPS) for the Controller Working Position, Version 6- 91/1, EUROCONTROL.

Hopkin, V. D. (1982), *Human Factors in Air Traffic Control*, AGARDograph No. 275.

Kerns, K. (1990), Data Link Communication Between Controllers and Pilots : A Review and Synthesis of the Simulation Literature, Mitre Corporation, MP-90W00027.

Kleinman, D.L. (1990) : Cooridnation in Human Teams: Thoeries, Data and Models. IFAC 11th Triennial WorldCongress, Tallinn, Estonia, USSR.

Majchrzak, A., Chang, T.C., Barfield, W., Eberts, R. and Salvendy, G. (1985). Human Aspects of Computer Aided Design.

NSTB (1976), *Human Error in Air Traffic Control*, National Transportations Safety Board, in Human Factors Bulletin, Vol 19, No 5, 1976.

Parsons H. M., (1985), Automation and the Individual: Comprehensive and Comparative Views, Human Factors, 27(1), pp. 99-112.

Sheridan, T. B., (1980), Computer Control and Human Alienation, technology Review, Vol. 83, (1), pp. 60-76.

Wiener, E. L. & Curry R. E. (1980), Flight-Deck Automation : Promises and Problems, Vol.23, No. 10.

Part 4
CRITICAL INCIDENT STRESS MANAGEMENT

18 Not only the sharp end: a flight attendant's viewpoint

Barbara Dunn, Aircraft Cabin Safety Specialist

Imagine if you will the following scenario.

You are sitting on the forward flight attendant seat just aft of the flight deck on a B-727. Your passengers are buckled in and your aircraft is ready to land. Without your knowledge there is a serious problem and the next thing you know you are thrown out on impact and you are found by the rescue personnel in a tree.

Among your numerous serious injuries are multiple compound fractures in both legs and you spend the next 4 ½ years in hospital.

Or perhaps you are winging your way home on a DC-9 when you notice smoke seeping out from underneath the rear washroom door. In a matter of minutes the cabin starts to fill up with smoke and very soon you are not able to see your hand in front of your face.

You know you have to sit down for landing but you can't see to find your seat. You try desperately to give your passengers evacuation instructions but the smoke is so thick and toxic you can't take a breath without wondering if it will be your last. Eventually, after what seems like an eternity, the DC-9 lands and you evacuate when you feel you can no longer stand the conditions in the cabin.

Neither of these stories are fiction. They are two of the many stories shared by the flight attendants with whom I have had the privilege of working over the past 17 years.

While not all of the stories are dramatic in nature, the survivors of the incidents all share one thing in common. They all experienced Post Traumatic Stress Disorder to some degree.

The definition of Post Traumatic Stress Disorder or Critical Incident Stress as it is sometimes called, is simple. It's the normal response to an abnormal situation or experience.

Thanks to the members of the cabin crew who have been willing to share their experiences and feelings with us over the years, we now know what to expect following an accident and we have developed some very effective ways of helping.

We have learned that the symptoms fall into 4 basic categories.

Psychological-Emotional Reactions

1. Psychic numbing, disbelief and bewilderment
2. Guilt
3. Fear vulnerability, powerlessness
4. Phobias
5. Mood Swings
6. Irritability and anger
7. Rage
8. Anxiety
9. Depression, loneliness
10. Sadness
11. Grief

Cognitive reactions

1. Temporarily impaired thought processes
2. Mental confusion
3. Inability to prioritize
4. Reduced trust in one's own judgement and decision-making ability
5. Shortened attention span
6. Impaired memory function
7. Limited creative ability
8. Difficulty with speech
9. Repetitious thoughts

Physiological reactions

1. Colds, flu
2. Gastro intestinal distress

3.   Skin eruptions and rashes, hair loss
4.   Changes in sexual energy, attitudes and behaviours
5.   Low energy level
6.   Chest pains, hyperventilation, hypertension
7.   Sleep disorders, nightmares, early morning awakening
8.   Head aches, allergies
9.   Menstrual changes
10. Muscle spasms, back pain
11. Grinding of teeth or clenching of jaws

Behavioral reactions

1.   Productive
2.   Non-productive

Productive behaviour

1.   Increasing awareness of self care; rest, nutrition, medical care, etc.
2.   Becoming increasingly comfortable with the expression of feelings
3.   Gathering accurate information about the disaster in response to his/her own need to understand and eventually integrate the experience, or
4.   Protecting oneself from information
5.   Solidifying and renewing friendships
6.   Spending time alone and with others according to his/her own needs
7.   Being willing to be self-honest and expressive of a multitude of feelings
8.   Becoming involved, when the time is right, in efforts to prevent further incidents, and in sharing knowledge for the benefit of others who may be confronted with a similar experience.

Non-productive behaviour

1.   Abuse of alcohol or self-medicating with other non-prescribed drugs. Misuse or unwarranted refusal to take prescribed medications and to otherwise cooperate with medical and/or psychological treatment
2.   Becoming excessively busy in a frantic effort to escape thoughts and feelings
3.   Displacing anger into self and onto inappropriate others. Jeopardizing relationships by provoking negative interactions
4.   Neglecting oneself in general health care and grooming

5. Forcing oneself to a premature return to anxiety producing duties in an effort to prove courage to self and others, and to avoid working through feelings.
6. Becoming accident prone, for example, being careless in safety measures - driving recklessly, falling, etc.
7. Withdrawing into isolation and refusing appropriate interaction with others
8. Engaging in suicidal preoccupations or behaviour, (any suicidal communication -- verbal or nonverbal, of high or low lethality risk -- must be professionally evaluated and responded to appropriately. Any suicidal conversation or action is a cry for help).
9. Spending money flamboyantly as if to avoid realities or buy peace of mind, making impulsive major life changes.

The strongest of the emotional responses seems to be guilt. Guilt for not saving passengers, guilt for being alive when a fellow crew member died, guilt for just being alive, guilt for not being able to help the investigators find the cause of the accident. No one said the guilt is rational, but it is very real.

Many of the symptoms listed can be lessened or totally alleviated if you have a proactive program involving, but not limited to a Critical Incident Stress Debriefing. The sooner you can get to your survivors, whether they be pilots or flight attendants and encourage them to express their feelings, the sooner they will start their recovery process.

The debriefing process is not complicated, but should not be attempted without appropriate training. Confidentiality is the cornerstone of the program and must never be eroded.

Other areas of concern are:
1. Salary continuance programs for survivors
2. Validation of experience involving the training department
3. Involvement and education for family members
4. Training programs for accident response personnel focusing on sensitivity training
5. Rehabilitation programs designed to retrain crew members who may not wish to return to flying
6. Scheduling flexibility in order to allow flight attendants to work as an additional crew member as necessary or stay away from a specific aircraft type.

It is important to remember that any interaction with a survivor will either help or hinder their recovery.

The symptoms may appear immediately or may take several weeks or months and after appearing they can be short or long term.

We have also learned that what may be traumatic to one crew member may be of little or no consequence to another. A crash is the most obvious trauma producing event but other incidents can also produce identical symptoms. These include, but are not limited to:

1. Hyjackings
2. In Flight Medical Emergency
3. Preparation for evacuation with uneventful landing
4. Violent weather incidents
5. A series of unrelated and seemingly minor incidents.

When dealing with survivors we must remember they are all individuals. Just as their reactions may differ, their recovery time will also differ greatly and patience is the order of the day.

In closing I would like to offer a list of Do's and Don'ts. These may also be helpful when dealing with next of kin.

Do's and don'ts for dealing with survivors/next of kin:

Do:

- Answer Questions as completely as possible
- Encourage expression of Feelings even if it's uncomfortable for you
- Listen
- Emphasize Confidentiality
- Remind them their reactions are normal
- Look after "creature" comforts
- Establish an open wallet policy

Don't:

- Tell them you know how THEY feel
- Tell them that they shouldn't feel what they are feeling
- Tell them they shouldn't feel guilty
- Tell them to stop crying
- Lie
- Make promises you can't keep
- Force them to talk.

Think of how you would like to be treated and respond in kind.

19   Debriefing British POWs after the Gulf War and released hsotages from Lebanon: lessons learnt for use in a wide variety of critical situations including aviation

Gordon Turnbull, BSc, MRCP, MRC Psych, FRGS
Clinical Director, Traumatic Stress Treatment Unit, Ticehurst
House Hospital, East Sussex, UK

The drawing together of the hitherto somewhat disparate concepts of traumatic stress reactions, battleshock, and combat stress reactions under the single mantle of Post-Traumatic Stress Disorder (PTSD) upon publication of DSM-III in 1980 Diagnostic and Statistical Manual of the American Psychiatric Association, had a considerable impact on the thinking of British psychiatrists. There were those who were frankly sceptical about the global nature of the conceptualisation and some believed that very little extra had been of value to what was already known. However, military psychiatrists in the UK were very interested in the utility of the framework. It arrived just in time because it was during the initial period of digestive assimilation that the Falklands conflict occurred.

The Falklands War was mainly a naval engagement and it was appropriate that the most intense interest in developing new treatment strategies for Combat Stress Reactions and establishing PTSD was demanded, of necessity, from the Royal Navy psychiatrists. Awareness spread by gradually increasing contact with the sucesses of this venture. The British Army needed to develop similar facilities to cope with the endemic demands of the conflict in Northern Ireland as well as their veterans from the Falklands War. The Royal Air Force was beginning to look at post-ejection phenomena and flying stresses in a different way in the light of the successful nature of PTSD treatments. It was the

Lockerbie Air Disaster just before Christmas of 1988 which concentrated Air Force minds. The aftermath of the air disaster involved about 1000 Air Force personnel. The RAF Mountain Rescue Teams who were rushed to the scene at the very beginning requested assistance in an entirely unprecedented way from the RAF Psychiatric Division. There was no question as to what the response to this request should be: it had to be affirmative. There was, however, the question of what to do. The team which responded to Lockerbie drew heavily upon the writings of Mitchell and Dyregrov and the concepts of CISD (Critical Incident Stress Debriefing). This was the birthplace for the debriefing techniques which were used in four settings during 1981. In sequence these were: released British POWs after the Gulf War; John McCarthy; Jackie Mann; Terry Waite. These four experiences were all unique occasions; all provided new lessons and demanded different things from the debriefing teams. The main lesson learnt was, without any doubt, that the approach to each situation had to be FLEXIBLE. There were also some general features which could be offered as lessons learnt.

Prisoners of war

The very first requirement, and it needs to be an absolute one with no blurring around the edges, is PERMISSION and SANCTION. This has to be at all levels of authority. The POWs will go along with an established practice as they emerge from their ordeal distressed, angry and looking for organised structure. Therefore the procedure to debrief before going home has to be an integral part of military protocol and not seen to be the experimental whim of psychiatrists. As a corollary to this, it is very important that the debriefing team has a clearly defined leader and that the overall management of the situation is by the psychiatrist. Physical medical checks are obviously of importance to the released POWs but they need one group to see them through all the elements of the debriefing, and I believe that this role falls naturally to the psychiatrist. In military situations this authority will usually have to be written down in the form of an operation order by the commander of high rank, since the psychiatric team has to have the authority to define time, place, readiness to go home and has to be free to use its own judgement. Of course authority has to be used with sensitivity and respect.

Emerging POWs need to be ISOLATED to begin with, in their group, to protect them from unnecessary invasions of privacy such as from the media. If a hospital environment has to be used during the debriefing, then steps should be taken to "de-institutionalise" the surroundings as far as possible; for example, window drapes, carpeting and pictures on the walls. Divan-type beds instead of customary hospital beds and the other items all serve to NORMALISE the situation for those released.

The POWs should be provided with personal items of clothing which they should be permitted to choose themselves. WRITING materials give a sense of freedom of self-expression. WATCHES are of great importance in the restoration of time orientation, self-control of time and movement and has huge symbolic value in that it is always one of the first items to be taken away by their captors and the watch has now been given back. It symbolises the potential of a future as the hands or the figures move inexorably on.

The psychological debriefing had two phases. The first was in answer to the question "WHAT HAPPENED TO YOU?" A new development was the discovery that not only was it convenient to combine this initial phase of the psychological debriefing with the intelligence debriefing but the manoeuvre provided mutual enhancement of both procedures. The ex-POW declared that he felt supported through the unpleasant business of recollection of events, comforted by the realisation that he would not have to repeat the story again, and that he could really concentrate on recalling everything because he would not have to do it again. The last factor, of course, received high approval by the military intelligence personnel. The COMBINATION OF PSYCHOLOGICAL AND INTELLIGENCE DEBRIEFING may prove to be the most important development to emerge from the process described in terms of future planning.

The second phase satisfied the question "WHAT DO YOU FEEL ABOUT WHAT HAPPENED TO YOU?" Somehow, setting about putting a layer of emotional meaning on top of cognitive understanding seemed to be acceptable, controlled and constructive. It seemed easier to recognise distortions this way. Misinterpretations were sorted out, the WHOLE experience integrated and comprehended which then, at least, opened up the opportunity to move into the future.

The ex-POWs were debriefed in GROUPS and care was taken to group together those veterans who had shared the most similar experiences. Two dedicated debriefers were assigned to each group and saw the process through with their group. All people involved met together for talks about post-traumatic stress reactions designed to enlighten about their potential emergence and to defuse anxieties about their significance as features of chronic, possibly life-long psychopathology. PTSD WAS NORMALISED in this way. Pamphlets had been designed to allow these messages to be taken home. One version of the pamphlets was designed to suit the ex-combatant and another aimed at informing the spouses and close relatives. There was also complete permission to make an individual approach to a debriefer outside the group. Of course any "hoarded" material which would be of value in the group setting was encouraged to be brought up again in group. Again, always with the permission of the individual concerned. All of these strategies were designed to restore personal dignity, integrity and independence, while maintaining a group identity.

The debriefers were debriefed and supported by SUPPORT DEBRIEFERS. We were all aware after the Lockerbie experience of the "RIPPLE EFFECT" of the disbribution of stress and the benefits of talking through impressions gained during the work with others not involved "at the coalface". These support debriefers must be trusted and respected and experiences known to all the debriefers. The role of the support debriefer was to listen, identify tiredness and over-involvement. Also to facilitate emotional ventilation among the debriefers. He also tested hypotheses and offered alternatives, encouraged and avoided directing strategy, always mindful of the importance of the leader-role for the team of debriefers and those being debriefed.

It did not prove possible to reunite the ex-POWs and their spouse or close relatives on this occasion, but it was the original intention to do so, and this would have met with the universal approval of the released prisoners. I think on reflection that the idea that the ideal environment for such a debriefing should be non-medical - because the medical environment represents an institution - is more theoretical than practical. There are distinct advantages in undertaking such a procedure in an environment which is associated with healing and fixing things. Practical advantages also would include the respect for privacy and the proximity to medical facilities.

I can report that the end result of the ten days which were spent in this way appeared to be sucessful. It is always difficult to measure the degree of success in such a venture but, if it can be regarded as objective evidence of success, there is no doubt that by the end of the debriefing the group had regained their group identity, had a better grasp of what they had been through, had restored a sense of humour, and most important of all, had decided that it had been better to have gone through this period of debriefing (despite the necessary delay in returning to loved ones) prior to going home rather than simply going home. There can be no greater approbation than this; no better customer approval.

Hostages

To a large extent the debriefing of the released POWs provided a template for the debriefing of the released hostages from Beirut. The formula asking the two cardinal questions of "WHAT HAPPENED TO YOU?" (to provide a cognitive structure for the events) and then "WHAT DO YOU FEEL ABOUT IT?" (to provide emotional meaning and understanding) seem to be equally appropriate in this different context.

One difference was that there was more time available for planning the debriefings. Even though the timescale was different in the three cases it still proved possible to meet the primary relatives before meeing the primary victim. In the case of the last hostage there was sufficient time to meet all of the primary relatives and actually acquaint them with the place in which the debriefing would take place (always provided that the released hostage would CONSENT with the plan) so that the only unknown factor was the state of the released hostage himself. In this way preparation involved collating information from as many sources as possible including the British Foreign and Commonwealth Office and Military Intelligence.

The team of primary debriefers was deliberately chosen to be sufficiently ample to provide a member for each primary victim or group of victims (such as children) and one liaison officer (Mr Fixit) who would deal with the administrative demands such as phone calls. Although all those involved were active Air Force personnel, only Mr Fixit was in uniform to provide a plausible INTERFACE with the authorities.

The debriefings of the three British released hostages, John McCarthy, Jackie Mann and Terry Waite, all took place in the secure environment of a Royal Air Force Base in the UK. The officers' mess accommodation afforded a suite of rooms which could be effectively cut off from the rest of the building. This allowed the released hostages to choose for themselves (which of course was effective therapy in itself) how much exposure they and their families wanted to be exposed to from the outside world and the media. Media interest was intense. It seemed from our previous experience that it would not be prudent to advise full exposure to the media too soon since the questions they were likely to ask would be invasive and reports might provide another person's, perhaps distorted, view of what the experience had been like. This was not a time for translation or sensationalism but a time for the released individuals to review their incarceration privately, taking as much time as they required, and to make sense of it for themselves. Once CONTROL had been re-established for themselves then there was ample opportunity for the world to know what happened.

We realised that the newly released hostages were not in a position or in sufficiently integrated mental state to be able to control an interview with the media. The temptation to debrief to "the world" rather than in private with family and the debriefing team would be enormous.

The work of the debriefers began long before the actual releases in the case of the second and third release. In the first case the only opportunity to get to know the family and other significant people was during the flight to Damascus which led to the pickup. However, this proved to be an invaluable time, and efforts were directed to get to know the close associates and families of the other two hostages. The debriefing work was a source of curiosity to the media during the first release. It also was a source of considerable irritation to them until they realised that there were sound reasons for gradual exposure to the media rather than flooding, and always under the control of the released hostage and his family themselves.

So much for the techniques, but what of the results? Since this was not designed to be a research project there are no objective data to assess the successfulness or otherwise of the debriefings. In the case of the released POWs after the debriefing there was an obvious return of group identity, a restoration of a sense of humour, and, perhaps most important of all as an indicator of a positive influence, the most antagonistic of those debriefed became the most supportive of the idea that the process should be adopted as a normal sequel to having been a POW and that the debriefing process should become an accepted part of rehabilitation and management after release. All of those managed in this way have returned to service life and fulfil their accustomed roles. Of course it could be that this would have happened anyway, nevertheless it seems that some at least of the difficulties experienced by these war veterans would have worked their way into the time during which they began the business of returning to family and professional lives, unless they have been given the opportunity to settle these issues during the assimilation period which the debriefing provided. In the case of the three Beirut ex-hostages, the debriefings provided an opportunity to integrate the experience and start to move into the future and there is some evidence that other released hostages from other nations have experienced difficulties in adjustment, which the British ex-hostages have not experienced.

The Royal Air Force debriefing teams have no doubts that it is a valuable course to follow. The evidence is accumulating that debriefing during the acute phase confers a positive advantage in avoiding the development of late-onset or chronic post-traumatic stress disorders and our experience has been strongly supportive of this belief.

The same, or very similar, principles could be applied to the critical ingredients which occur in the world of aviation. Aircrew tend to be emotionally controlled individuals who might, predictably, suppress traumatic stress reactions. This is a perilous situatation for them as the unprocessed (or relatively underprocessed) reactions tend to produce long-term symptoms which compromise both professional and personal performance. The underlying biological basis to long-term PTSD has been well-identified and can persist for many years, leading to both psychological and physical illness and premature medical requirements.

20 Critical incident response program: a Canadian perspective

Capt. D.R. Andersen, Canadian Air Line Pilots' Association

Introduction

The Human Performance Division of the Canadian Air Line Pilots Association recently completed development of a Critical Incident Response Program (CIRP). Several events during the past few years had indicated to us that there was a need for this type of program in order to better serve the requirements of our pilot membership in this area. As you may well understand, developing and implementing a program of this nature on a national scale does not happen overnight, and the implementation of the program within our member airlines is still ongoing.

CALPA Human Performance Division

At the present time, the Human Performance Division consists of four working committees or groups. Only two of these committees, the Aeromedical Committee and the Pilot Assistance Committee are relevant to the CIR Program, so I will attempt to describe them and their current functions within the HPD in some detail prior to discussing the CIR Program itself.

The Aeromedical Committee presently consists of approximately 35 volunteer pilots who attempt to keep as current as possible on the ever changing medical licensing requirements within Canada, as well as the latest developments in medicine which may have a bearing on our pilot group. These pilots provide a medical referral service for CALPA member pilots who request assistance in dealing with the more traditional aeromedical problems or who are having licensing difficulty due to medical problems. The Aeromedical Committee maintains a number of medical specialists on retainer who are used quite extensively by our membership through this referral service. The Aeromedical Committee volunteers are all required to attend an initial three day Peer Helper course as part of their training prior to becoming involved actively within the committee. I will get back to this course shortly as it is relevant to our Critical Incident Response Program (CIRP).

The second committee or group within the HPD which is relevant to the CIR Program is the Pilot Assistance Committee. It, along with the Aeromedical Committee, comprised the Pilot Health Division until early last year. The Pilot Health Division was then reorganised with the inclusion of two former Tech/Safety committees, the Human Factors Committee and the Training and Licensing Committee, into the current Human Performance Division.

Pilot Assistance Committee

The Pilot Assistance Committee is, I believe, unique to CALPA and since it is the foundation upon which our CIRP is based, I will attempt to explain its history and development in more detail. This committee is comprised at the present time of approximately 85 pilot volunteers from across all member airlines. These pilots volunteer their time and services to assist fellow pilots who may find themselves in need of some assistance or direction in order to again function normally in their personal or professional lives.

The committee can trace its roots back to the early 1960's when some of the World War Two pilots who had joined Trans Canada Airlines after the war started exhibiting the now well recognised symptoms of alcohol addiction. Back in those days, Management had neither the knowledge nor the programs necessary to deal with these problems in a constructive manner. In other words, dismissal was the order of the day if repeated warnings didn't work. Recognising that there had to be a better way to deal with these problems, several line pilots formed an unofficial and, essentially, an underground organisation, to try and help alcoholic pilots deal with their problems before Management were forced to act.

130

This small, underground committee was so successful that it has now, thirty years later, expanded to the point where it is formally recognised by all CALPA member airlines and deals with a wide range of potentially career threatening problems that may affect a pilot. Each member airline has trained pilot volunteers on the Committee. Within most member airlines there are a number of official programs in place, sanctioned by CALPA, the airline and, where required, Civil Aviation Medicine, to deal with several areas of mutual concern. The Critical Incident Response Program recently became the latest of these to be accepted by most of our member airlines.

These Pilot Assistance volunteers presently represent approximately 70% of the total "workforce" within the CALPA Human Performance Division. They provide confidential assistance and, when required, professional referral service to any pilot who approaches them for help, regardless of the problem. In fact, alcohol related problems make up a relatively small portion of their actual activities today. Their methods differ from the traditional EAP approach to this type of service in that they may become **proactive** if it is felt that a situation warrants it. In other words, if it comes to their attention that one of our pilots appears to be having difficulty with an issue in his/her life, they do not always wait for the pilot to approach them, but will often intervene and offer whatever assistance may be appropriate. This has become an accepted way of doing business within our pilot group.

All Pilot Assistance and Aeromedical Committee volunteer members are required to take a Basic Peer Helper Training Course prior to becoming involved in Committee work. This three day course essentially helps an individual to better understand him/herself and teaches or reinforces basic listening and communication skills. The course provides those of us responsible for the Committee's activities with a method of screening volunteers in order to ensure some standardisation among the members and the services provided. It also helps identify individuals who may be effective in more specialised roles within the Committee.

The Committee members are provided with a list of professional resources within their geographical area and, as well, several medical specialists and professional counsellors are kept on retainer by the HPD. These professionals are used in cases where a Pilot Assistant encounters a situation where a sympathetic ear and some good feed back are not sufficient to rectify the problem. Many of our Committee members eventually volunteer to take additional training in other areas to enable them to be more competent in dealing with specialised cases. The latest specialised training to be offered is in critical incident stress response.

Development of the Critical
Incident Response Program

Nearly ten years ago, it was recognised that a program of this nature could be beneficial to the pilot group, but at that time it was felt that the expertise required to develop such a program would have to come from sources outside CALPA. It became evident that the cost of developing a program utilising outside resources was prohibitive and, reluctantly, the project was dropped.

Approximately three years ago, however, in the spring of 1991, the demand for such a program resurfaced. We have had several major accidents and incidents involving member airlines during the six or seven preceding years. As a result, we reconsidered our previous assessment of the resources required to establish a Critical Incident Response Program and elected to utilise those already available to us, that is, our own Committee members and the professional resources maintained on retainer.

We initially had eight volunteer Aeromedical and Pilot Assistance volunteers and our primary mental health resource professional participate in, and audit, the Critical Incident Stress Debriefing courses being presented by Dr. Geoffrey Mitchell through the Critical Incident Stress Foundation. These committee members, in concert with the Chairman of the HPD, Captain Dave Noble, then determined that it was feasible to proceed with the development of the CIR Program within CALPA.

In order to obtain some feedback from members who had been involved in critical incidents, Mr. Brian Murray, our primary professional resource in this area, interviewed on videotape the crews from the Air Canada Cincinatti DC-9 accident and the Gimli B-767 accident, as well as several other crew members who had been involved in events that would qualify as critical incidents. These interviews served two purposes. First, it provided us with some first hand information about what these crew members felt would have been beneficial for them in the way of assistance after the accident. Secondly, with the crews' permission, these video tapes are to be used during CIRP training sessions to give our volunteers some insight into the post traumatic stress problems which may be encountered after events such as these.

All pilots and professionals who had been involved in the initial planning stages then met for a three day workshop retreat and developed the document which now serves as our 'Manual of Policy and Procedures' for Critical Incident Response.

Our initial intent had been to develop a program which dealt mainly with aircraft accidents and incidents, However, it became apparent to us that these situations comprised only a small portion of what could be considered critical incidents in a pilot's life. As a result, we attempted to develop a program which would address different levels of stress, always keeping in mind the desirability of maintaining the lowest level of intervention possible to deal with the situation. The result is our classification of three different categories of critical incident stress events and the different approach to be taken, at least initially, in each category.

Categories

The highest level of response would be for an event which we classified as Category A. This would include any accident or incident resulting in fatality or serious injury and/or serious damage to an aircraft or property. Response by a full CIRP Team including a mental health professional would be mandatory to an event in this category. A defusing would be done if it could be accomplished within the appropriate time frame and a formal debriefing would be mandatory for all crew members involved. Follow up services would also be provided for an event in this category.

The second level, or a Category B event, would include any accident or incident resulting in some aircraft or property damage, or an event which had the potential, under slightly different circumstances, to have become a Category A. CIRP Team response to this category would normally consist of an intervention by peer member(s), using a modified debriefing process while keeping open the possibility of going to a formal debriefing if the situation warranted. Again, follow up services would be provided if required.

Category C events would be those, other than accidents or incidents, which could increase the pilot's level of stress or threaten his/her license. Examples of this type of event would include promotion or upgrade failures, disciplinary proceedings or aeromedical problems. Normally, initial contact for these events would be a low level peer intervention, during which a determination would be made on whether to move to a higher level of response. Much of this type of intervention is already being performed by the Pilot Assistance members and has proven to be very effective over the years.

It should be noted that the perception of severity of the event as held by the pilot(s) involved can raise the required response accordingly. That assessment will be made by the senior CIRP team member present, in consultation with the Program's Mental Health Coordinator.

133

Structure

The Critical Incident Response Program is provided under the direction of the CALPA Human Performance Division in cooperation with airline management. The HPD provides a budget for the operation of the Program and also provides any professional and clerical resources required to ensure that the program functions in an efficient manner.

The Critical Incident Response Program is the responsibility of an HPD-appointed Program Coordinator. The Program Coordinator, in conjunction with an HPD-appointed Mental Health Coordinator is responsible for the day to day operation of the program, including the promotion of the program, selection and training of team members and quality assurance of all team members. They report directly to the Chairman of the Pilot Assistance Group at the present time, who is in turn responsible to the Chairman of the HPD.

Each airline has a Company Team Coordinator who is responsible for that airlines selection of suitable pilots to be trained as peer support personnel. Each Company Coordinator acts as a liaison between the Program Coordinator and the peer support personnel and is also responsible for coordinating any team response required within his/her airline.

Conclusion

Although we are still in the early stages of implementation of this program within our member airlines, we feel that we have moved a long way in a fairly short period of time. Obviously, developing and implementing a Program such as CIRP is not a simple task. We were fortunate in that, due to the existence of the Pilot Assistance and Aeromedical Committees, we had in place two essential elements necessary to develop and implement a Program in a relatively short period of time. First, we had a nationwide organisation with a proven structure and good lines of communication. Secondly, we had a group of approximately 120 volunteer pilots who had already been trained in basic peer helping techniques and who were obviously interested in helping their fellow pilot. We therefore had no shortage of volunteers to train for this program, as many of them viewed it as an extension of the type of volunteer work they were already doing. We believe that using a peer based program will provide the most effective support for our pilot group and would suggest that any other organisation planning to set up a CIR Program seriously consider that approach as well.

21 The establishment of the Delta–ALPA Critical Incident Response Program

Capt. Alan D. Campbell, Air Line Pilots' Association

Numerous people have assisted in the development of the Delta-ALPA Critical Incident Response Program (CIRP) including Captain Dale Andersen of the Canadian Air Line Pilots Association (CALPA), and Greg Janelle, of Air British Columbia (BC). One of the most enjoyable aspects of working on this program has been the opportunity to know some very talented and generous people interested in this field. A number of airlines outside the U.S., including Finnair have already developed programs.

Several years ago I heard a presentation from Captain Mimi Tompkins. As Mimi shared her story I was dismayed that her experience with post traumatic stress had gone unidentified. Her sharing that story led to my interest and commitment to this program.

Capt. Lloyd Sauls Chairman of the Delta-ALPA Human Performance Committee is interested in establishing this program at Delta and he asked me to direct the project. As a result on February 17, 1994, the Delta MEC delivered the CIRP, at approximately 2:00PM, in New Orleans, La. The new infant was approximately three pages long, but required a gestation period of over a year and one half.

This presentation is an overview of the Delta-ALPA CIRP development as approved by the pilot group. Support is now being sought from Delta corporate and such time the program will fully implemented.

Research

An accurate and thorough information base is crucially important before development of a program plan. Numerous months were spent researching and familiarizing myself with stress, stress components, and ways to deal with post traumatic stress. Access to the Psychology Literature Review (a database of published psychological literature of the past twenty years) was invaluable and allowed me to review all empirical psychologically related studies. This review supported my thoughts about the necessity for this type of program.

My wife Linda, a licensed psychologist, introduced me to psychologist who had trained in disaster response. These affiliations allowed me access to further information, knowledge of training, and service going on within the mental health professions.

During this research I decided that the Critical Incident Stress Debriefing (CISD) model advocated by Dr. Jeffrey Mitchell, would best meet the needs of our situation. Dr. Mitchell's model was developed for ambulance, fire, and police services; however, I felt that with some minor changes it would meet the support needs of airline personnel.

The name CALPA used seemed much less emotionally laden than CISD so I adopted the term Critical Incident Response Program. I have diligently endeavored to remove emotionally laden words from the program description and have sought to develop a frame of reference designed not to trigger adverse response within the pilot community.

Review of other programs

I reviewed other programs within my airline that might overlap or focus on similar concerns. Professional Standards was one of the areas I reviewed. I attended an initial training program and gained a number of insights on organization and structure. It was there I realized the necessity to provide legal support for critical incident team members. Another assistance program called Human Intervention Management Strategy (HIMS) provides resources to impaired pilots. I intend to interact with that group prior to completing this program.

Establish the needs and scope of the program

There continues to be no standard within the U.S. by which psychological service needs are identified and provided for the flight crew or other airline personnel (Williams, Solomon, & Bartone, 1988). There is a growing awareness that psychological services are crucial to the short term coping and long term adjustment of those involved in an incident. Dr. Donald Hudson (Senior Aeromedical Advisor to US-ALPA) has stated that without this type of support, 60% of all pilots involved in critical incidents will not be in the cockpit two years after the accident or incident. The potential uniqueness of the needs of the airline crew remains unacknowledged.

Defining the scope of this program has been quite difficult and time consuming. Within ALPA there has been considerable concern about providing too wide a scope of services (ie. duration, frequency, and inclusion of family members). Long term mental health support is considered outside the scope of this program. When long term needs are identified, services will be provided by referral to an appropriate mental health provider.

Initially our program will limit service to the following areas:
1. Members who are involved in an aircraft accident.
2. Members who are involved in an aircraft incident that results in injury to persons or damage to equipment.
3. Members who are involved in an on-the-job incident where there exists the real or perceived threat of death or serious injury to a crew member or passenger.
4. Members and staff who have participated in aircraft accident investigation activities.
5. CIR team members, would receive assistance when necessary.
6. Limited assistance will be provided the spouse or immediate family members of crew members involved with a critical incident or accident.

This does not represent the full scope of necessary services, but is a compromise in order to begin these services.

Develop a plan

Most important is the development of a strategy for accomplishing the task. A fundamental aspect of any plan is the clear and understandable explanation of the program. This includes education and successful solicitation of support from those individuals in the decision-making loop. I developed a paper that has served to introduce and educate the reader to the importance of establishing a CIR Program. It was necessary to anticipate resistance, address stated concerns, and negotiate resolution to the problem. This program was coordinated and discussed with all the key players. As problems and concerns arose they were addressed and changes were incorporated within the program. Resistance was expected and a lot of time was spent in selling the program to those making the decision to implement or not.

The Delta-ALPA CIR Program today

This brings us to the Delta CIR Program today. The pilot group has initiated

137

the program, and is negotiating Delta corporate acceptance. Thus far the program has been discussed within the safety and flight operations area and has been well received. Presently I am seeking to justify the program on a cost/benefit basis. Costs in training, medical expenses, and workmen's compensation that result from stress related experiences should be significantly reduced. Benefits may include the fostering of employee goodwill, a reduction in employee absenteeism, and alleviation of stress related health problems including abuse of alcohol and cigarette use, ulcers, migraine headaches, and colitis.

It is hoped that a company wide program will result. Pilot peers would serve the pilot group as well as a part of the Delta CIR Team. Initial discussions suggest that a program of wider scope than initially presented may be desired by Delta corporate. I hope that this program will be launched as possible.

At inception we intend to use licensed psychologists and psychiatrists with a specialty in emotional trauma and who can utilize the Mitchell model. We wanted to provide the highest level of confidentiality of any information the flight crew might share. In the U.S. licensed psychologists and psychiatrists have privileged communication (a legal status) rather than confidentiality (an ethical expectation). We know this will be somewhat limiting for us and as we gain more experience this requirement may change, particularly as we begin to learn the capabilities of available mental health providers and seek to balance those against the cost of this program. There is considerable interest in this type of program among U.S. and European airlines.

The CIR team will consist of a peer volunteer at the defusing and a mental health provider and several peer volunteers at debriefings peer volunteers will have a detailed application and review process. There will be initial training of approximately three days and ongoing recurrent training and advanced skills training. They will be recommended and accepted by fellow peers and have no history of corporate problems. A high level of integrity will be required as well as empathetic qualities. Peer volunteers are important because they are less stigmatizing than other support resources.

Information concerning normal reactions of crew members involved in a critical incident will be provided the spouse or immediate family. Education is considered basic to this program. All involved are provided information concerning post accident or incident emotional trauma, stress reduction techniques, and the CIRP.

This program is not reportable on the FAA medical. Should long term assistance be required that may be reportable but may not affect flight status. The CIR Program is quite active in the U.S. today. In February 1994, Federal Express hosted the first U.S. airline CISD conference. It was attended by representatives from Federal Express, Alaska, Aloha, Continental, Delta, Northwest, United, and Air British Columbia. These air carriers are actively building critical incident response programs. American Airlines is operating a company led program within the E.A.P. area.

Critical Incident Stress Debriefing (CISD)

I would like to share a short overview concerning CISD. It was introduced by Dr. Jeff Mitchell in 1983. CISD refers specifically to the Mitchell model. It was initially designed to address stress issues of personnel within the police and fire services. People in high stress avocations are often unaware of unusual stress reactions because they are used to "normal" stress and have a harder time identifying "abnormal" levels of stress.

A critical incident is any event impactful enough to produce strong emotional reactions in people now or later. CISD is not psychotherapy or counseling in the strict sense of the word. It differs primarily because it includes a teaching component. It teaches about stress reaction. It also teaches stress survival techniques. The goal of CISD is to mitigate the impact of the critical incident. It serves to assist the involved personnel, accident investigators and the family. The CISD team makeup is approximately one-third mental health professionals. They have backgrounds in dealing with emotional trauma and have led defusings and debriefings using the Mitchell model.

Shared characteristics of the CISD Model and the CIRP Model include the roles of confidentiality, defusing, and debriefing: (a) Confidentiality is of a major concern. The only record keeping would be to evaluate program effectiveness and team member involvement. Immediacy is of great importance, (b) The defusing should take place within three to five hours but certainly not more than twenty-four hours maximum. The defusing gives an opportunity to ventilate feelings and consists of introduction, exploration, and education phases, (c) The debriefing is necessary if no defusing occurred or the incident was particularly powerful. It normally would be one to three hours and include a mental health provider and several peer volunteers.

This program is a win for all involved. The Air Line Pilots Association would be providing a service to its membership and demonstrating a valuable support function. The corporation gains improved employee morale, reduced sick leave, and maintains a valued employee, who has increased satisfaction with management.

Part 5
ERROR ANALYSIS

22 A methodological framework for root cause analysis of human errors

M. Pedrali, Centre National de la Recherche Scientifique, ARAMIIHS, Toulouse, France
G. Cojazzi, Commission of the European Community, Joint Research Centre, Institute for System Engineering and Informatics, Ispra, Italy

Introduction

The design of modern plants and systems has considerably changed in the last few years, owing to few but significant occurred accidents. In particular, those accidents that risked the population safety have stressed the increasing role played by humans in the event management, due to errors or erroneous behaviours performed by operators. The complexity of human performances added to their great variability, with respect to system performance, has been generally accepted as one of the main difficulties of the human factor analysis. This has led to develop methodologies whose goal is the analysis of accidents or incidents involving human factors (Leplat and Rasmussen, 1987). This analysis can be conducted in a *retrospective* or in a *prospective* way. In the first case, root causes of erroneous behaviours are found within a given sequence; in the second case, likely manifestations and consequences of erroneous behaviours can be analysed starting from possible causes. Root Cause Analysis (RCA) methodologies are conceived for the restrospective study of accidental or incidental sequences involving human actions with the specific goal of identifying those ones entailing errors.

The methodology that we propose for the RCA allows the analyst to detect firstly what are the incorrect behaviours performed by the operator(s), and then, to identify the possible causes that made that erroneous sequence to begin and to progress. The analysis of human actions is based on a model or paradigm of human behaviour and on a classification or taxonomy for human errors. The use of the paradigm is particularly important to detect the causes of cognitive errors, whereas the use of a taxonomy is essential to clearly distinguish between internal and external causes affecting human performances. The employed taxonomy is completely coupled with the paradigm of human behaviour and it clearly distinguishes between the manifestations and the causes of an inappropriate behaviour.

The RCA methodology was preliminary applied to an aeronautical accident: the Alitalia disaster in Zurich, 1990 (Cacciabue et al., 1993). The application was made while the methodology was not completely formalized; in this paper the complete theoretical formalization of the methodology for the retrospective analysis of incidents/accidents will be presented.

The proposed RCA methodology

The methodology is subdivided in two parts, namely: Erroneous Actions Identification (EAI), and Causal Analysis (CA). In the first part (EAI), the analyst is capable to reconstruct the incident/accident and to detect what were the human errors performed; these errors are, indeed, the manifestations of erroneous behaviours. Subsequently (CA), the analyst finds in which phase of the cognitive model the error arose, and what was/were the triggering cause(s). The reason of this subdivision is of methodological nature: incorrect actions need firstly to be singled out within the context before looking for their causes.

Erroneous Actions Identification (EAI)

The first part of the methodology aims at identifying those particular actions that are considered not "justifiable" and for that reason are considered erroneous; the meaning of this adjective (justifiable) will be clarified in the following. Those actions are the error's manifestations (the final result of a cognitive process), whose causes will be searched by means of the CA. The Erroneous Actions Identification is not only preparatory to the Causal Analysis, but it is also the main reference when performing the CA. The EAI is further subdivided in three steps, namely: Data Collection; Event Time Lines; Action Detection.

Data Collection. In this phase all the material concerning the overall evolution of the incident/accident is gathered. It is important to collect information regarding not only the actions the operator(s) accomplished, but also the data available on instrumentations, signals, cues. Sources of data can be audio/video recordings, incident forms, elicitations of operators and experts, and so on.

144

Event Time Lines. This step aims at giving a time oriented representation as detailed as possible of the event; information collected in the previous phase are now structured in a working sheet as to better detect performed actions. These filled sheets also provide a structured illustration of all the gathered material. The execution of this step is very important since it offers a clear picture of the accident/incident and it will constitute the main reference in the search for the causes of erroneous actions. This concept was originally proposed by Pew (Pew et al., 1981) and it can be suitable employed in any accident analysis.

Action Detection. "From the point of view of accident analysis, of highly proceduralized man-machine systems, it can be reasonably said that each deviation from the procedures, written or even tacit ones, is a potential error" (Cojazzi and Pinola, 1993). At this point of the RCA if possible, all the actions performed by the operator that do not correspond in general to what the procedure foresees have to be detected. This is a very crucial phase of the method since, for instance, a procedure may not always be very well detailed or, the operator may have found himself to make a sort of compromise following two or more parallel procedures or even, in certain cases, the operator has to really invent a new one.

The detection of the erroneous actions is pursued by the aid of an expert; the sequence of actions reconstructed by the Event Time Lines is compared with the procedure the operator has to follow - the procedure itself ought to be conceived as a step by step description of tasks & goals in order to make the comparison easier. From the comparison, those actions that did not respect the procedure can be subdivided into **justifiable** actions, and **unjustifiable** actions. The distinction is necessary because actions that are not usually envisaged by the procedures, are sometimes unavoidable owing to particular circumstances. These particular circumstances can be exceptional weather conditions for example or, more often, an inadequate functioning or a failure of an instrument. However, only those actions that cannot be "justified" are the *erroneous actions*, in terms of error manifestations or modes.

Causal Analysis (CA)

In the second part of the methodology, the causes of the human erroneous actions, previously identified, are ascertained by an appropriate taxonomy that was originally proposed by Hollnagel (Hollnagel and Cacciabue, 1991) and that is described here with reference to an updated version (Cacciabue et al., 1993). Causes can be divided in two categories, namely: **System-related causes** (external causes), **Person-related causes** (internal causes). The System-related causes refer to the conditions that are external to the operator and that can be ascertained in an objective sense, whereas the Person-related causes refer to internal conditions that cannot be directly observed and that are typical of the cognitive process. The System-related causes play the role only as the possible external conditions that may set off or modify a **Person**-related cause. However, the Person-related causes can be the origin of an incorrect action, that is, a System-related cause is not always necessary to give rise to an erroneous behaviour.

The taxonomy is conceived to allow a safety analyst to find the internal and external cause(s) of an inappropriate behaviour by starting from its *manifestation*, called *phenotypes*. Four tables, in relation with the four cognitive functions of the cognition (Observation, Interpretation, Planning, Execution), are tied each other by cause-effect links. The reference cognitive model, namely the Simple Model of Cognition (SMoC), is schematically represented in figure 1 where the four cognitive functions previously mentioned together with the envisaged connections are well characterized.

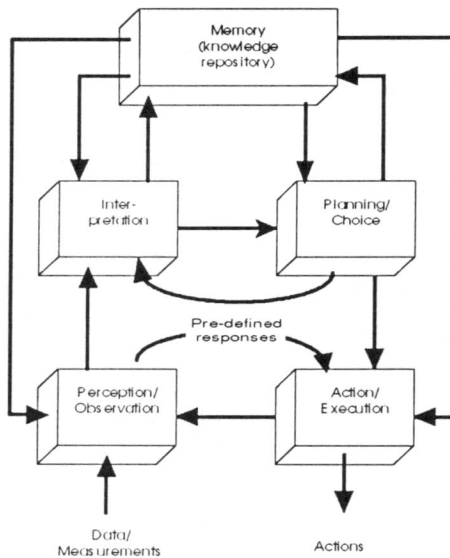

Figure 1 - The SMoC (Hollnagel & Cacciabue, 1991).

The table of the taxonomy related to phase of the Execution is structured according to the following template.

General Effect	Specific Effect	General Cause	Specific Cause
Phenotypes	*Phenotypes*	$\Rightarrow$ to *General Effect* *(PLAN)*	*System Related Cause*

In the *General Effect* and *Specific Effect* columns there are the Phenotypes or Error Modes; in the other two columns (*General Cause, Specific Cause*) there are respectively the Person Related Causes of the Execution, that link this phase with the Planning phase, and the System Related Causes responsible of some of the listed phenotypes. The phenotypes contained in the Specific

146

Effect column better identify what are the contents in the General Effect column. The tables of the taxonomy related to phases of Planning, Interpretation and Observation are, on the contrary, respectively structured according to the following template. In the *General Effect* and *General Cause* columns there are those Person Related Causes that link the different cognitive phases in a causal-effect connection: what is "cause" in a phase, in its turn, it is "effect" in the previous. In the other two columns (*Specific Effect*, *Specific Cause*) there are respectively the Person Related Causes of a particular phase of the SMoC (Specific Functions), and the Person Related Causes not tied to a specific phase of the cognitive loop (General Functions). The Person Related Causes contained in the Specific Effect column better identify what are the contents in the General Effect column.

General Effect	Specific Effect	General Cause	Specific Cause
from ⇒ *General Cause* *(EXE/PLAN/INT)*	*Specific Function*	⇒ *to* *General Effect* *(INT/OBS)*	*General Function*

Taxonomy Application. In the second part of the methodology (CA), the first step is the identification of the erroneous behaviours with the **phenotypes** listed in the Execution table of the taxonomy. The cause(s) of erroneous behaviour must be searched inside the tables by covering the four phases of the SMoC, from the Execution up to the Observation, and moving from left **(General Effect)** to right **(Specific Cause)** inside the tables.

When the phenotype is identified, its cause may be found in the column "General Cause" or "Specific Cause" of the Execution table. If the analysis has been postponed to another one, it continues from the column "**General Effect**". In the corresponding cell of the "Specific Effect" column there are the possible effects of that phase whose causes can be found:

1. In the "**General Cause**" column. In this case, the analysis continues in the previous phase; therefore, causes must be searched in the previous levels of the process of cognition.

2. In the "**Specific Cause**" column. In this case, column, there may be two other possibilities:

   - One or more System Related Cause(s) could set off or modify the Person Related Cause(s) reported in the "Specific Cause" column (general function).
   - One of these general functions was self triggered.

3. In the "**Specific Cause**" and in the "**General Cause**" columns. In this case there is the "branching": the analysis continues in the previous phase but one or more cause(s) are detected also in the current phase, in the "Specific Cause" column.

147

4. In the "**System Related Cause**" table. One or more System Related Cause(s) could set off or modify the Person Related Cause reported in the "Specific Effect" column.

The analysis is led in this way in all the tables until the root causes of the phenotype are found.

Conclusion

The technological development has shifted the role of operators of complex systems more towards supervisory tasks. This augments the difficulty to analyse and understand complex situations in which operators played a role in the initiation and in the progression of an accident. A suitable methodology is needed to attack this thematic; a framework for the assessment of the root causes of human erroneous actions is presented in this paper. The method is based on two main steps that aim at the identifications of the actions that deviate from the procedures (EAI) and at the identification of the causes, internal or external, that affected the human performance. The envisaged methodology is quite general and might in principle fruitfully applied to many domains; however, its practical application to a specific domain would require a fine tuning of the human error taxonomy that is at the basis of the method.

This work was partially funded by ALENIA (contract No B 9364018) in the frame of an ESA contract on Human Dependability. The authors wishes to thank P.Carlo Cacciabue and E. Hollnagel for many interesting and fruitful discussions.

Reference

Cacciabue, P.C., Pedrali, M., Hollnagel, E., (1993), Taxonomy and Models for Human Factors Analysis of Interactive System: An Application to Flight Safety, Proceedings of the 2nd ICAO Flight Safety and Human Factors Symposium, 12-15 April 1993, Washington DC, USA.

Cojazzi, G., Pinola, L., (1993), Root Cause Analysis Methodologies: Trends and Needs, To be presented at *PSAM II*, 20-24 March 1994, San Diego, CA, USA

Leplat, J., Rasmussen, J., (1987), 'Analysis of human errors in industrial incidents and accidents for improvement of work safety', in Rasmussen et al., *New Technology and Human Error*, J. Wiley.

Hollnagel, E., Cacciabue, P.C. (1991). Cognitive Modelling in System Simulation, Proceedings of Third European Conference on Cognitive Science Approaches to Process Control, 2-6 September 1991, Cardiff, UK.

Pew, R.W., Miller, P.C., Feeher, C.E., (1981), Evaluation of Proposed Control Room Improvements through Analysis of Critical Operators Decision, EPRI-NP 1982,891.

23 Simultaneous error: during altitude deviations

Capt. Ron Raymond, Ansett New Zealand
Dr Anne R. Isaac, University of Otago, New Zealand

Summary:

The 1992 FAA Altitude Deviation Study included an altitude awareness programme for pilots. The programme is almost identical to systems employed in Australia and New Zealand where carries are still experiencing altitude busts. Although we are familiar with error triggers, their effect on cross monitoring receives little attention. Both pilots are vulnerable; a development conducive to error on one hand and inadequate monitoring on the other. The Altitude Deviation Study published by the FAA in 1992 mentioned approximately 1,160 altitude deviations which were attributed to persons other than flight crew alone suggesting the involvement of air traffic services. It was the intention of this investigation to review data concerned with altitude deviations by both air crew and air traffic control in New Zealand. It was also hoped to establish whether these incidents could be categorised in terms of degraded monitoring and attributable to the phenomenon of simultaneous error in either or both aviation environments. Finally recommendations will be made.

Situational awareness is often lost during smultaneous error incidents. However, as recent CFIT accidents illustrate, crews often lack time to recognise situational awareness cues. Firstly we will look at pilot error and then at ATC statistics and attempt to focus on some of the similarities which occur using altitude deviations as examples.

149

From recent observations on the flight deck of a series of Ansett flight (6 sector) throughout NZ, a number of errors which could fall into the simultaneous category were recorded. There were 18 errors in total. These errors occurred during the first and last 15 minutes of the flying sectors.
- Flapsetting/AC pump/Landing gear/Airbrakes
- Readback/Hearback - delayed, slow or none
- Altitude - ascent/descent
- Navigation
- Cabin calls

The recent altitude bust study by the FAA (1993) (in conjunction with US air and ALPA) were able to categorise these errors into four main types; Information processing, Decision making, Task priority and Malfunction. When analyzing the present pilot errors into these categories a similar tend emerges.

Table 1 Pilot errors by sector and type

Sector	Information Processing	Decision Making	Errors * Task Priority	Malfunction	Totals
1	1	1	1	1	4
2	1	1	1	0	3
3	2	1	0	0	3
4	1	1	1	0	3
5	3	1	2	0	6
6	1	0	0	0	1
Totals	9	5	5	1	20

* Each error could be classified by several types

As with the FAA study information processing seems to be the most common error type. If this is a repeatable observation then it may implicate the standard operating procedures for pilots and ATC's as technology increases the monitoring role and therefore the information processing system.

When analysing altitude deviations from ATC in NZ over the last three years similar trends emerge. ATS incidents are divided into nine categories and the incidents regarding altitude deviations may fall into many of these categories. Results indicate as co-ordination, communication, instruction and information problems have decreased, procedural problems have increased.

150

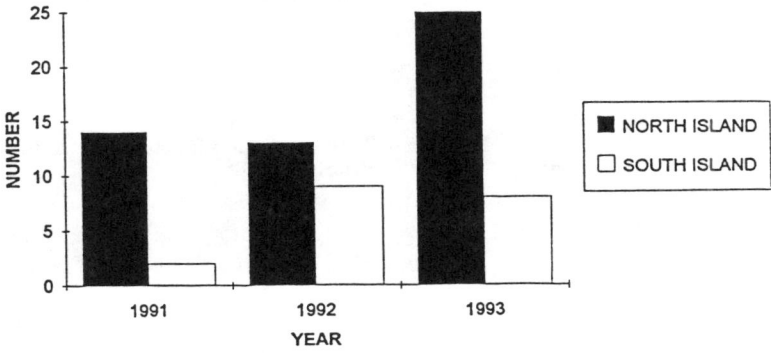

Figure 1 New Zealand altitude deviations 7/3/91-31/10/93

There were a total of 71 reported altitude deviations from 1991 (7/3/91) to 1993 (31/10/93). (Figure 1)

The number of aircraft movements during this time was approximately 3.2 million (91:892, 371: 92: 1,116,897: 93:1,191,021).

Without figures for November and December 1993 it is difficult to predict the annual high points although figures to date would suggest March, June and July as possible problem months. A breakdown of the altitude by pairs deviation suggest, as other studies have, that the most frequent pair is 10,000 - 11,000 feet (Figure 2).

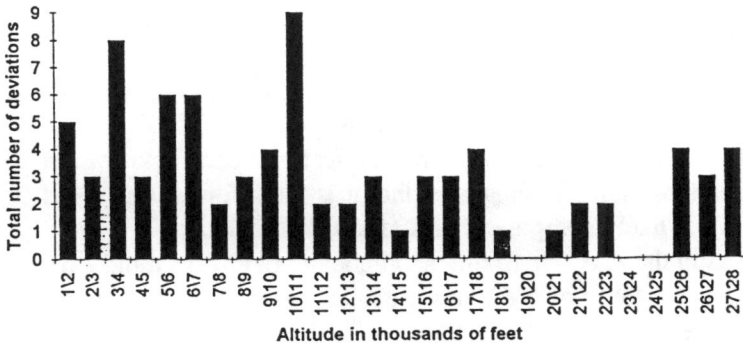

Figure 2 Number of altitude deviations by altitude pairing

The higher numbers in the lower altitudes reflects the numbers of light aircraft included in this graph. Most other research only includes medium and heavy commercial aircraft.

If these altitudes deviations are classified in the same way as the pilot errors the following trend emerges.

151

Table 2 ATC reported altitude deviations by error type and task

Task	Information Processing	Errors * Decision Making	Task Priority	Malfunction	Totals
Formulate & issue clearance	3	1	0	0	4
Monitor/Process clearance	27	14	10	1	52
Hearback	11	0	0	0	11
Implement clearance & monitor	9	1	14	0	24
TOTALS	50	16	24	1	91

* Each error could be classified by several types

As with the pilot errors the main problems lie in the information processing stage and in these ATC reports during monitor/ process clearance tasks. The other task which indicated a large number of errors was to implement clearance and monitor compliance and was usually caused by task priority, although information processing was again implicated.

In both these situations, pilot error and altitude deviations from ATC, the errors are found most commonly in information processing.

The phenomenon of simultaneous error on the flight deck is known in single and multicrew situations and it is suggested that it is the information processor system which is the most likely component of a task to be degraded.

The problems in the ATC environment, although not always multi - controller, are not dissimilar and a study of NZ altitude deviations again implicates problems with information processing as causal factors.

The information processing system is known to be degraded by such factors as task overload, preoccupation, inexperience, boredom, unfamiliar situations and surprise, leading to decisions of risk acceptance.

Faith in cross - monitoring, readback and hearback is often misplaced: stress often effect pilots and ATC's simultaneously. Even unaffected operators can be led into a state of error or uncertainty by confused colleagues.

The flightdeck and to some degree the ATC environment, can be degraded nine ways the resultant loss will take three forms, risk acceptance, loss from the loop, or high stress level.

These nine situations can be listed as follows:
Mutual confidence	-	risk acceptance
Underestimating risk	-	risk acceptance
Task overload	-	out of loop
Boredom	-	out of loop
Preoccupation	-	out of loop
Inexperience	-	out of loop
Task overload	-	stress level
Unfamiliar situations	-	stress level
Surprise	-	stress level

The following examples are illustrated from a pilot prospective in New Zealand

Mutual confidence:
Most NZ airline crews have received some form of HF training. Pilots lacking this training have at least heard of CRM. Training standards are high and check procedures rigorous. Crew members, in our small industry, quickly become aware of each others limitations - friendly trust is a New Zealand/Australian ethos. I also believe that this attitude can generate negative problems - misplaced confidence on the flight deck: acceptance of decisions, superficial cross monitoring and tolerance of flight deviations.

Underestimating Risk:
Airline pilots rarely crash due to lack of fundamental skill. However, faulty decisions can lead pilots into conditions beyond their ability or the aircraft's capability. Covert pressure also exists on line crews - a decision to persist is often the easiest option.
Mutual confidence and faulty risk assessment often arise in combination, occasionally appearing as a breakdown in situational awareness.

Task Overload:
Task overload can remove support pilots from 'the loop' and effectively destroy their cross monitoring function. Emergency checklists, flight management systems, airways clearances and abnormal situations can destroy the normal work balance. An error by the operating pilot can remain undetected.

Boredom:
Boredom/low arousal is often evident in both pilots. Not only is the operating pilot susceptible to mistakes, support monitoring also breaks down.

Preoccupation:

Preoccupation occurs on the flight deck when pilots are put into an abnormal situation, amended clearances, the unusual that requires prompt attention. If the support pilot cannot solve the problem quickly there is a risk the operation pilot will also become involved and the scene is set for both a mistake and failure to detect the mistake.

Inexperience:

Inexperience produces various errors - stress, omission, commission and cognitive overload are some examples. It is possible, even likely, that both pilots will become overloaded leading to mutual errors at a time when monitoring has broken down.

Surprise:

Depending upon the circumstances, an impulsive decision can lead to major error. A simple combat instruction ' Break right-Go' posses no problem as no decision is involved. However an unexpected instruction to enter a different hold or non-precision approach, without time to plan its operational needs, degrades situational awareness and increases workload. Although the time needed to ensure normal cognitive flow is surprisingly brief, it is never the less essential. It is evident that this is a significant factor in erroneous RTO decisions and a number of CFIT accidents.

The recent altitude bust study by the FAA emphasised the value of procedure oriented SOP's : in this case an altimetry procedure that significantly reduced the frequency of altitude busts. Despite this change however, a limited number of busts continued to occur, reflecting the Australian and NZ experience, where the procedures have existed in airline SOP's for some time. Evidence from ATC altitude deviation statistics suggest that as with the pilot error data the main area of concern seems to be in information processing by those concerned and evidence further suggests that simultaneous errors can occur in this domain in both multi-crew and ATC units.

Further research is under way in NZ regarding these problems.

References

Pope, J.A. (1993) Research identifies common errors behind altitude deviations. *Flight Safety Digest* June.

Raymond R. (1990) Simultaneous cockpit errors. Paper presented to Aviation Psychology Symposium Massey University.

U.S.F.A.A. (1992) Altitude Deviation Study: A descriptive analysis of Pilot and Controller Incidents. Unpublished document. October.

24 Controlled flight into terrain avoidance: why the ground proximity warning system is too little, too late

William H. Corwin, PhD, Honeywell Technology Center, USA

This paper identifies two classes of Controlled Flight Into Terrain (CFIT) accidents, enroute versus approach & landing, in order to determine the information requirements needed to avoid the CFIT event. Historic examples, based on accident reports, will be used to delineate underlying common "causes" of these two classes of CFIT accidents. A thorough understanding of the unfolding events in both classes of CFIT accidents will be used to illustrate the insufficient protection provided by existing radar altimetry based warning systems. Borrowing terminology from the Medical model, the Ground Proximity Warning System (GPWS) can be viewed as treating symptoms (insufficient terrain clearance) as opposed to the underlying malady, in this case information deficits which cause an event. Information deficits that contribute to CFIT are addressed in sufficient detail so that breakdowns in information processing can be addressed at more than one stage of processing. Additionally, other viable technologies are addressed that can remedy the information deficit problem, thereby lengthening the alerting time to the flight crew which can increase the probability of CFIT avoidance.

Controlled flight-into-terrain

In order to ensure a common basis of understanding the following definition of Controlled Flight-Into-Terrain (CFIT) will be adopted in this paper. The following CFIT Definition is borrowed from Wiener (1977):

> Controlled flight into terrain accidents are those in which an aircraft, under the control of the crew, is flown into terrain (or water) with no prior awareness on the part of the crew of the impending disaster.

In response to the loss of life as a result of CFIT accidents, a system was developed to alert the flight crew to unsafe terrain clearance. The Ground Proximity Warning System (GPWS) was developed by the Sundstrand Company. The GPWS utilizes radar altimetry to warn pilots of unsafe terrain clearance, unsafe terrain closure rate, and unsafe descents below the glideslope or shortly after takeoff. The following table outlines the various modes of the GPWS that alert different unsafe terrain proximity conditions.

GPWS Mode		Basic Equipment		Advanced Equipment	
		Alert	Warning	Alert	Warning
1. Excessive descent rate		**	"Whoop Whoop Pull Up"	"Sink Rate"	"Whoop Whoop Pull Up"
2. Excessive terrain closure rate		**	"Whoop Whoop Pull Up"	"Terrain Terrain"	"Whoop Whoop Pull Up"
3. Altitude loss after takeoff or go around		**	"Whoop Whoop Pull Up"	"Don't Sink"	"Whoop Whoop Pull Up"
4. Unsafe terrain clearance while not in the landing configuration	4A Proximity to terrain –Gear not locked down	**	"Whoop Whoop Pull Up"	"Too Low Gear"	"Whoop Whoop Pull Up"
	4B Proximity to terrain –Flaps not in a landing position	**	"Whoop Whoop Pull Up"	"Too Low Flaps"	"Whoop Whoop Pull Up"
5. Descent below glideslope		"Glideslope"	**	"Glideslope"	**
6. Descent below minimums		**	**	"Minimums"	**

Figure 1   Modes and alerts of the ground proximity warning system

Over the years this system has undoubtedly alerted flight crews to the immediate danger of terrain proximity, and in the process saved many lives. The limitation to GPWS is that it treats an outcome (the *symptom*) namely unsafe terrain proximity or closure, it does not address the cause of how the flight crew allowed this unsafe condition to develop (the information deficit *disease*). If the flight crew knew of the developing unsafe condition they would rectify it before the GPWS annunciation of imminent peril. In effect, GPWS is attempting to break the last link in the causal chain of events leading to an accident. A better solution would be to foil the development of the chain of events leading to unsafe terrain proximity.

The purpose of this paper is to address the information deficits themselves that are contributory to the CFIT accident. The intent here is not new, in fact this paper follows a time-honored tradition with notable examples provided by Ruffell-Smith (1968), Wiener (1977), Loomis & Porter (1981), and Bateman (1991). This paper will deviate from the previous approaches by attempting to use accident reports to identify similar causes of CFIT accidents from an operational standpoint.

Information deficit

All CFIT accidents are not caused by the same information deficit. Clearly, at the end of the line, the flight crew were unaware of the impending impact, but the antecedents may be different. In order to systematically address the information deficit issue it makes sense to address CFIT based initially upon phase of flight. The same information deficit can be responsible for a CFIT

156

accident in various phases of flight, but the phase of flight distinction makes conceptualization of the information deficit problem easier. There are two obvious classes of CFIT based upon the phase of flight distinction: Enroute (level flight)-the aircraft is holding altitude; Descent, Approach, and Landing-the aircraft is intentionally trying to make ground contact (land). Below is a list of accident examples representative of both classes of CFIT accident. The key information deficit is denoted in *italics*.

Enroute (level flight)

Lateral path error
- Mt. Erebus (Air New Zealand), all 257 lives lost.
  An incorrect flight plan was correctly programmed into INS. The aircraft descended in "white-out" conditions in an attempt to provide better viewing conditions during this sight-seeing tour of Antarctica. The lateral displacement error (approx. 20 nm) was sufficient to place the flight path directly into an extinct volcano, Mt. Erebus.

It is important to discriminate between precision and non-precision approaches because of the reliance on *bad* flight director data being implicated as a contributing factor to the accident in the precision approach CFIT class of accidents.

Descent/approach/landing (precision approach)

Autopilot/flight director mode confusion
- Boston (Delta), all 89 lives lost.
  During a "rushed" ILS approach, the flight crew selected the Go Around mode of the Flight Director, instead of ILS. The split-cue flight director provided amplified lateral tracking of the localizer and wings-level pitch guidance. The aircraft collided with the seawall demarking the airport boundary (165 ft right of centerline, 3,000 ft short of RWY 4R).
- Kansas City (Buffalo Air), all 5 lives lost (freighter).
  During an ILS approach in severe rain, the flight crew (Pilot Flying - Captain & Pilot Not Flying - First Officer) had difficulty communicating height above ground level versus barometric altitude. In addition, suspect Flight Director guidance was responsible for ground contact 3 nm short of the airport (RWY 1).
- San Francisco (JAL), no lives lost.
  During an ILS approach in fog, the flight crew (Pilot Flying - Captain & Pilot Not Flying - First Officer) had difficulty intercepting the localizer and began a premature descent. An ATC call for traffic distracted the flight crew during the critical moments of the premature descent. The aircraft "landed" in San Francisco bay 2.5 nm short of the airport (RWY 28L). The passengers and aircraft survived the water landing, the aircraft was in fact returned to service.

Descent/approach/landing (non-precision approach)

Autopilot/flight director mode confusion
- La Bloss Hill (Air Inter), 87 of 96 lives lost.
  During a VOR approach, over high ground the flight crew selected a
  Vertical Speed descent mode inadvertently. Instead of selecting a
  Flight Path Angle of -3.3° the Pilot Flying (Captain) selected a descent
  rate of 3,300 feet per minute. (The Mode Control Panel has a dual
  function knob, V/S and FPA.) At the accelerated descent rate (3,300
  ft/min instead of ~1,500 ft/min) the aircraft collided with a ridge 7 nm
  DME from the aerodrome VOR.
- Bangalore (Air India), most lives lost.
  During a visual approach, the flight crew had selected an Open
  Descent mode with the Flight Director in the off position, the
  consequence being that the aircraft would not level off at the selected
  altitude. The aircraft collided with the ground in sight of the airport.

Mental picture of path is flawed
- Unalakleet (Mark Air), no lives lost.
  During a VOR/DME approach, the Pilot Flying (Captain) developed a
  "mental picture" of an interim descent point at 10 miles DME (as
  opposed to the correct waypoint at 5 nm DME). This error, although
  spoken for cross-check purposes, went unchallenged by the Pilot Not
  Flying. At 10 nm DME the Captain descended to the 5 nm DME
  clearance altitude and made contact shortly thereafter.
- Berryville (TWA), 92 lives lost.
  During an VOR - DME approach, the flight crew was "cleared" for
  the approach by the air traffic controller. The crew interpreted this to
  be an immediate clearance to the next level-off altitude.
  Misunderstanding of the intent of the clearance coupled with
  confusing graphics on the approach plate led the crew to assume that
  is was safe to let down. At 25 nm DME the aircraft collided with a
  ridge, the crew was flying at the altitude appropriate for 6 nm DME.

It is important to note that GPWS is not always successful in saving the day.
The DC-10 involved in the Mt. Erebus CFIT accident was GPWS equipped.
Without a doubt, the flight path of the doomed aircraft was not in the intended
design envelope for GPWS, but the unsuccessful save is more of a result of
addressing the information deficit issue when flight conditions had become *in
extremis*. The aircraft was to fly a low-level sight-seeing tour of Antarctica.
On the day of the flight, the crew was dispatched with an erroneous set of
waypoints which were dutifully (and correctly!) programmed into the aircraft
INS system. The crew let down into marginal visual conditions because they
believed they were over McMurdo Sound, and believed their current heading
would take them over the low terrain of McMurdo Ice Shelf. Instead the 20
nm discrepancy (the *true* information deficit!) in the flight plan brought them
directly over Ross Island. At 1,500 feet MSL the aircraft traveling at ~250
KIAS contacted the gently rising slope of Mt. Erebus. The GPWS had simply

been overcome by the closure rate of the swiftly moving aircraft. Subsequent simulation tests of pilot performance revealed the following:

> The performance of the GPWS was evaluated and it was assessed that the warning was in accordance with the expected performance in the "terrain closure" and "flight below 500 feet without flaps and undercarriage extended" modes of the equipment (modes 2A and 4 respectively). The profile of the terrain prior to the impact was reconstructed in Air New Zealand's DC-10 simulator and the performance of the aircraft was evaluated to determine if the collision could have been avoided in response to the warning and that the warning was in fact given at the maximum time before impact that could be expected. The flights in the simulator indicated that experienced pilots would **not** have avoided a collision and that the warning given was in accordance with the design specifications of the GPWS. (page 23 of Aircraft Accident report No. 79-139)

Addressing the information deficit problem

Path error

The advent of the map feature for the NAV display in an electronic flight instrument system has reduced the guess-work involved with position location relative to the planned flight path. What remains is to represent the flight path on a display with a terrain-based reference. This can be accomplished using elevation (topographic) data or cultural features (e.g., aeronautical charts). The digital map feature available in military aircraft is specifically designed to provide the pilot with lateral position awareness. The lateral plan view can be augmented with a depiction of the vertical situation, which may be a more compelling display if terrain data is depicted on the same display as the programmed and actual flight path.

Autopilot/flight director mode confusion

The most sinister information deficit is that involving the selection of guidance modes. It is clear that the Pilot Flying can get "tunnel vision" on the compensatory tracking task of following a Flight Director. The presentation of raw data deviation information from the localizer & glideslope and/or mode annunciation are **not** attended to because of attentional tunneling on the tracking task involving the Flight Director. What can be done? Certainly a standardization of an optimized mode annunciation norm would remove the intra-manufacturer idiosyncrasies that currently plague the cross-qualified pilot. But to address flight operations in existing fleets it may be wise to adopt a procedure developed by USAir to avoid altitude deviations. On USAir flights the Pilot Flying points to the altitude window of the mode control panel after selecting a new altitude, the Pilot Not Flying must then point at the selected altitude to affirm that they have attended to the new selection. A case can be made that overt acknowledgment of current and armed mode would alleviate the confusion now experienced by ill-fated crews. Of course this is an empirical question that warrants a controlled investigation.

References

Anon: Japan Airlines Douglas DC8-62, JA 8032, 22 November 1968, San Francisco Bay. Washington, National Transportation Safety Board file no. A-0002, 1969.

Anon: Delta Airlines DC9-31, N975NE, Boston, MA, 31 July 1973. Washington, National Transportation Safety Board report no. NTSB-AAR-74-3, 1974.

Anon: Trans World Airlines Boeing 727-231, N54328, Berryville, VA, 12 December 1974. Washington, National Transportation Safety Board report no. NTSB-AAR-75-16, 1975.

Anon: Air New Zealand DC-10, ZK-NZP, Ross Island, Antarctica, 28 November 1979. Office of Air Accidents Investigation, Ministry of Transport, 1980.

Anon: Buffalo Airways Boeing 707-351C, N144SP, Kansas City, MO, 13 April 1987. Washington, National Transportation Safety Board report no. NTSB/AAR-89/01/SUM, 1989.

Anon: Markair, Inc. Boeing 737-2X6C, N670MA, Unalakleet, Alaska, 2 June 1990. Washington, national Transportation Safety Board report no. NTSB/AAR-91/02, 1991.

Anon: Air Inter Airbus A320 F-GGED, 20 January 1992, Mont Sainte-Odile (Bas Rhin), Paris: Ministry of Planning, Housing, Transport and Space, 1992.

Bateman, D (1991). How to Terrain-proof the World's Airline Fleet. Presented at the Flight Safety Foundation 44th IASS, Singapore.

Loomis, J. P. and Porter, R. F. (1981). The Performance of Warning Systems in Avoiding Controlled Flight-Into-Terrain (CFIT) Accidents. Presented at the First International Aviation Psychology Symposium, 20-22 April 1981, Columbus, OH.

Porter, R. F. and Loomis, J. P. (1981). An Investigation of Reports of Controlled Flight Toward Terrain (CFTT). Draft Report prepared under Contract No. NAS2-10060 for NASA Ames Research Center.

Ruffell-Smith, H. P. (1968). Some Human Factors of Aircraft Accidents Involving Collision with High Ground. *The Journal of the Institute of Navigation*, **21**, 354-363.

Wiener, E. L. (1977). Controlled Flight into Terrain Accidents: System-Induced Errors. *Human Factors*, **19**(2), 171-181.

25 Visual perception of object motion and depth: implications for poor visibility

A.H. Reinhardt-Rutland, Psychology Department, University of Ulster at Jordanstown

Failure in perception provides a likely complete or partial explanation in a significant proportion of aviation accidents, whether on the ground or in the air (Hawkins, 1987). Potential failures in perception are many; see Boff et al (1986) for a comprehensive review. The purpose of this paper is to outline theoretical and empirical evidence relating to one likely source of failure for which there has been considerable recent laboratory research - the interpretation of visual motion at the retina and its effect on perception of object motion and motion-based depth.

It is argued that relative visual motion is of prime importance in the context of unaided operation of all types of powered vehicles, since classical sources of kinaesthetic information - which complement visual motion - are attenuated or misleading. Implications concern perception of object motion and motion-based depth in poor visibility: reduced sensitivity and/or non-veridicality may be important. One application concerns conspicuity and lighting on the ground: attention to good general lighting may be more important than attention to conspicuity of individual objects. However, often alleviation may not be open to such manipulations: since perception is the most phenomenally-immediate of psychological functions - "seeing is believing" - education may be necessary.

Motion perception and kinaesthetic information

In principle, perception of object motion could be based on visual motion of the object's image across the retina, assuming the eye is entirely static. However, this rarely, if ever, holds: visual motion must be integrated with eye motion. Under the special conditions of ocular pursuit, motion should be signalled by kinaesthetic information associated with eye, head and body motion; visual motion of the tracked object occurs when tracking is inexact.

Such considerations suggest that motion perception in general can be based on comparison of retinal image motion with kinaesthetic information: motion is perceived when the two types of information do not match (Wertheim, 1981). Two observations support this theory. First, moving the eye with a finger leads to perceived motion of the visual scene; imposed, passive eye motion is not monitored by the visual system, so perceived motion must be largely conveyed by visual motion. Second, an afterimage created by staring briefly at a bright light source is perceived to move as the eye moves; the afterimage is fixed with respect to the retina, so perceived motion must be largely conveyed by eye motion (Mack & Bachant, 1969).

Relative visual motion and perceiving object motion and distance

However, such an account of motion perception has limited applicability in the context of powered vehicles, such as aircraft: kinaesthetic information is grossly attenuated, because the vehicle's means of propulsion substitutes for the observer's active motion. Yet object motion is still adequately perceived *in general* from within powered vehicles. The reason resides in *relative visual motion*: moving objects are normally perceived against a background of other objects. The importance of this is demonstrated in experiments in which one or two dots are viewed in an otherwise unstructured visual field: the threshold for perceiving veridically the motion of a single dot alone in the visual field is high, but if a second static dot is introduced - the dot's motion is now relative to this second dot - the threshold is dramatically reduced (Shaffer & Wallach, 1966). This has been confirmed for absolute and relative motion of large-scale patterns (Snowden, 1991). A further indicator of the importance of relative motion is the autokinetic effect: a single static dot in an otherwise unstructured visual field is perceived to move randomly after viewing it for a while (Boff et al, 1986).

The effect of relative visual motion is also pertinent to the relative depth cue of motion parallax - visual motion of static objects viewed by a moving observer varies with object-observer distance. Motion parallax is obvious when viewing out of the side window of a vehicle, but it equally applies to forward motion, except, in principle, in the centre of vision. That the observer's motion can lead to adequate perception of relative depth is now well-established (Rogers and Graham, 1979; Reinhardt-Rutland, 1993a,b).

Problems of motion perception while travelling in a powered vehicle are therefore likely during poor visibility; incidentally, the latter might result both from weather conditions and from windshield problems. Attenuation of visible background removes relative visual motion, thus entailing poor motion and depth perception of a sensed object. An example is borrowed from the roadway literature, which can be a useful source of evidence because of the relatively enormous number of casualties entailed on the road (de Kroes and Stoop, 1993): aids to conspicuity on the road - particularly retroreflective material and fog lamps - have not had a demonstrably great effect in reducing casualties (Shinar, 1985; Sivak, 1979). While retroreflective material can lead to detection of the user - perhaps a pedestrian - at distances well beyond the driver's required stopping distance, the environment *around* the pedestrian is unaffected: the driver still must perceive the pedestrian's motion and distance when relative visual motion may be sparse. As support, Polus and Katz (1978) found that street lighting, which affects conspicuity of pedestrian *and* environment, is effective in reducing pedestrian casualties. In a different context, many UK police authorities now regard road-side lighting as a priority for motorways (Reinhardt-Rutland, 1991): compulsory high-intensity rear lamps have had little clear influence on casualty rates, presumably again because their effect is mainly confined to increasing conspicuity of the user. Such evidence has obvious implications with regard to ground manoeuvres of aircraft and support vehicles.

Depth perception and known properties of objects

As an aside from the main issue in this paper, evidence suggests that motion parallax and other motion-based information for depth can be secondary to known properties of objects - particularly regarding relative size (Reinhardt-Rutland, 1993a, b); depth of moving objects is also affected (DeLucia, 1991).

Such evidence indicates a difficulty in isolating specific processing, such as that relating to visual motion: often a particular perceptual property can be conveyed in more than one way. It is unfortunate that information such as relative size can be important in depth perception, since it is easily rendered non-veridical: borrowing again from the roadway literature, this assertion might explain the over-involvement of small automobiles in road accidents in North America at a time when the average size of automobiles was reducing (Eberts and MacMillan, 1985). This has application to aircraft and support vehicles on the ground: perception of their distances may be partly dependent on their known sizes, leading to possible problems if new vehicles of unfamiliar size are introduced.

Visual motion contrast

Relative visual motion appears to be so important for perception that mechanisms in the visual system exaggerate it, often at the expense of veridically perceiving absolute motion. At the phenomenal level, this is reflected in the illusory motion of the moon against clouds on a windy night. The salient object need not be static: a moving object appears to move fast if surrounding objects move slowly and slowly if surrounding objects move fast (Loomis & Nakayama, 1973). Such examples of *visual motion contrast* are reflected in motion aftereffects, which are indicators of sensory processing (Reinhardt-Rutland, 1987, 1988); at the physiological level, analogous activity is reported in motion-sensitive neurons (Hammond, Mouat, & Smith, 1986; Mandl, 1985). More recently, simultaneous motion contrast has been linked to the "vista paradox": the observer's motion towards a stationary distant feature - perhaps a group of houses - viewed behind a rise in the ground or other feature obscuring the nearby landscape may elicit a perception of the houses moving away from the observer. The houses, because of their great distance, are virtually static in visual terms; the environment up to the obscuring feature elicits the most salient part of the observer's visual "flow-field" (Gibson, 1979), which in turn elicits motion contrast of the houses (Walker, Rupich, & Powell, 1989; Reinhardt-Rutland, 1990).

Simultaneous contrast can also affect at least one aspect of depth perception during observer motion: perceived slant-in-depth of a surface may be altered because of adjoining surfaces' slants (Graham & Rogers, 1982). Contrast, aftereffects and analogous neural activity are general to perception, affecting brightness, orientation, colour and size, as well as motion (e.g., Coren & Girgus, 1978). In common with other forms of contrast (Coren & Girgus, 1978), motion contrast can be elicited in the laboratory with highly impoverished stimuli, suggesting that it may be salient in conditions of poor visibility (Reinhardt-Rutland, 1985); other candidate conditions might therefore include featureless runways for aircraft landing or taking off.

Concluding remarks

Since perception is the most phenomenally-immediate aspect of psychological functioning, we tend to take perceptual experience for granted, unless there is compelling reason for doubting its veridicality. Supplementing this assertion is a belief that perceptual functioning is reasonably uniform across the population: except perhaps for relatively "low-level" or physiologically-based functions such as colour blindness or acuity, perception is often not amenable to obvious screening processes as are, say, motivational or personality factors. In the case of road vehicles, at least, the role of perceptual screening has generated much debate and conflict about possible perceptual functions that should be explored - but consensus is often lacking (Gale, 1991). Finally, it is of note that perception has been the most difficult of human information-processing functions to simulate computationally: while "high-level" functions - as exemplified by chess-playing - can be simulated proficiently, even seemingly basic functions of perception - such as segrating visual scenes into individual objects - have presented immense challenges, an assertion that becomes even more salient in considering visual motion (Ginsberg, 1993); a likely contributory factor to this seeming paradox is the above-implied lack of insight humans often have about their perception.

One implication that emerges from the above analysis concerns the importance of general ground lighting - if this can be achieved - entailing reduced reliance on conspicuity of individual items. More generally, Hawkins (1985) argues that aircrew should be educated to the range of possible failures in perception, so that they know under what conditions they may require supplementary information.

References

Boff, K. R., Kaufman, L., & Thomas, J. P. (1986), *Handbook of Perception and Human Performance,* New York: Wiley.

Coren, S., & Girgus, J. S. (1978), *Seeing is Deceiving*, Hillsdale NJ: Erlbaum.

DeLucia, P. R. (1991), Pictorial and motion-based information for depth perception, *Journal of Experimental Psychology: Human Perception and Performance, 17* 738-748.

de Kroes, J. L., & Stoop, J. A. (1993), *Safety and Transportation*, Delft: University of Delft.

Ebert, R. E. and MacMillan, A. G. (1985), Misperception of small cars, in: Eberts, R. E. and Eberts, C. G. (Eds.), *Trends in Ergonomics/Human Facctors II*, Amsterdam: North Holland.

Gale, A. G. (1991), *Vision in Vehicles - III*, Amsterdam: North Holland.

Gibson, J. J. (1979), *The Ecological Approach to Visual Perception*, Boston: Houghton-Mifflin.

Ginsberg, M. (1993), *Essentials of Artificial Intelligence*, San Mateo CA: Morgan Kaufman.

Graham, M., & Rogers, B. (1982), Simultaneous and successive contrast effects in the perception of depth from motion-parallax and stereoscopic information, *Perception, 11*, 247-262.

Hammond, P., Mouat, G. S. V., & Smith, A. T. (1986), Motion aftereffects in cat striate cortex elicited by moving texture, *Vision Research, 26*, 1055-1060.

Hawkins, F. H. (1987), *Human Factors in Flight,* Brookfield VM: Gower.

Loomis, J. M., & Nakayama, K. (1973), A velocity analogue of brightness contrast, *Perception, 2*, 425-428.

Mack, A., & Bachant, J. (1969), Perceived movement of the afterimage during eye movements, *Perception & Psychophysics, 6*, 379-384.

Mandl, G. (1985), Responses of visual cells in cat superior colliculus to relative pattern movement, *Vision Research, 25*, 267-281.

Polus, A., & Katz A. (1978), An analysis of night-time pedestrian accidents of specially illuminated crosswalk, *Accident Analysis & Prevention, 10*, 223-228.

Reinhardt-Rutland, A. H. (1985), A new visual factor in certain driving problems". In D. J. Oborne (Ed.), *Contemporary Ergonomics 1985* (pp. 60-69), London: Taylor & Francis.

Reinhardt-Rutland, A. H. (1987), Aftereffect of visual movement - the role of relative movement: a review, *Current Psychological Research and Reviews, 6*, 275-288.

Reinhardt-Rutland, A. H. (1988), Induced movement in the visual modality,*Psychological Bulletin, 103*, 57-71.

Reinhardt-Rutland, A. H. (1990), The vista paradox: is the effect partly explained by induced movement? *Perception & Psychophysics, 47*, 95-96.

Reinhardt-Rutland, A. H. (1991), Driving in poor visibility: Limits of perceiving motion, *The Police Journal, 64*, 22-25.

Reinhardt-Rutland, A. H. (1993a), Detecting slant-in-depth of real trapezoidal and rectangular surfaces: moving-monocular viewing equivalent to stationary-binocular viewing, *Journal of General Psychology, 120,* 177-185.

Reinhardt-Rutland, A. H. (1993b), Perceiving surface orientation: pictorial information based on rectangularity can be overriden during observer motion, *Perception, 22*, 335-341.

Rogers, B., & Graham, M. (1979), Motion parallax as an independent cue for depth perception, *Perception, 8*, 125-134.

Shaffer, O., & Wallach, H. (1966), Extent-of-motion thresholds under subject-relative and object-relative conditions, *Perception & Psychophysics, 1*, 447-451.

Shinar, D. (1985), The effects of expectancy, clothing reflectance, and detection criterion on nighttime pedestrian visibility, *Human Factors, 27*, 327-333.

Sivak, M. (1979), A review of literature on nighttime conspicuity and effects of retroreflectorization, *HSRI Research Review (University of Michigan), 10(3)*, 9-17.

Snowden, R. J. (1992), Sensitivity to relative and absolute motion,*Perception, 21*, 563-568.

Walker, J. T., Rupich, R. C., & Powell, J. L. (1989), The vista paradox: a natural visual illusion, *Perception & Psychophysics, 45*, 43-48.

Wertheim, A. H. (1981), On the relativity of perceived motion, *Acta Psychologica, 48*, 97-110.

Part 6
FEAR OF FLYING

26 Fear of flying: an investigation into aerophobia and its treatment

Maeve Byrne-Crangle, PhD, Aer Lingus Irish Airlines, Dublin Airport

A phobia is defined by Marks (1978) as a special kind of fear which is disproportionate to the demands of the situation and which cannot be explained or reasoned away. It is beyond voluntary control and invariably leads to avoidance of the feared situation.

The inability to travel by air is perceived by many aerophobics as a serious handicap. As a result of this problem, many restrictions are imposed on the quality of their lives, in terms of their ability to pursue business development, career promotion and opportunities, recreational pursuits such as holidays and cultural and educational interests.

The present research was conducted in two main parts. Study 1 consisted of an exploratory investigation into fear of flying into two samples of air travellers. Study 2 consisted of a comparison of the effectiveness of group treatments designed to overcome fear of flying.

Study 1

Study 1 involved two groups who were members of the public. The first group (N=104) perceived the experience of flying as non-threatening and non-stressful, and acted as controls. The experimental group (N=100) perceived the same situation as extremely stress-inducing and were classified as fearful flyers. This group self- referred themselves for treatment in a fear of flying programme.

In this Study fear of flying was evaluated by the following measures:-

1) Flight Stress History
2) Flight Stress Symptoms Checklist
3) Aeroanxiety Survey.

The findings from these measures showed many differences between the two groups which are as follows.

Results

Flight Stress History

A majority of fearful respondents (58%) perceived air travel as a dangerous mode of transport. Forty one percent indicated a response of severe anxiety while 51% indicated panic at the prospect of taking a flight.

The fearful subjects had an array of fears in addition to their air travel anxiety. The most common was a fear of heights which was reported by over half, followed by a fear of confined places (43%) elevators (37%) underground trains (28%) and tunnels (22%). A significantly smaller proportion of controls also reported fear in these situations, with one-third reporting a fear of heights.

A substantial proportion of fearful flyers (65%) resorted to alcohol and tranquillizers prior to confrontation with the phobic situation with the same high proportion continuing to consume alcohol throughout the flight. These methods had very little success in reducing their air travel anxiety. The control group had little or no need to resort to these substances.

Flight Stress Symptoms Checklist

Anxiety symptoms relating to cognitive, behavioural and physiological responses are frequently experienced by fearful air travellers prior to taking a flight, which for many continue to increase when actually in the flight situation. The results showed that the control group did not experience any elevation in stress levels either pre-flight or during flight.

Aeroanxiety Survey

In the Aeroanxiety Survey subjects rated their anxiety on a scale from 1 - 4 (with 1 indicating no anxiety and 4 indicating extreme anxiety) with regard to 57 flight-related situations.

The greatest fears were experienced when in the flight situation. Extreme anxiety was reported by 82% at the prospect of flying through turbulence followed by stormy weather (80%), crashing and dying (75%), engine noises (71%) and not having any control in the situation (68%). On the other hand the findings showed that 16% of the controls responded with extreme anxiety at the prospect of crashing and dying, while a mere 1% reported severe anxiety in response to the other situations.

Support for some of these findings comes from Howard, Murphy and Clarke (1983) and Walder, McCracken, Herbert, James and Brewitt (1987). In general it would seem from these studies that the major fears associated with flying consist of, heights, confinement , and crashing.

Discussion

When considering these results and those of the present study it seems that fear of flying cannot be considered a unitary phenomenon; it consists of a cluster of fears which are essentially part of the flight situation. The fears involved may be experienced singly or in combination and surface when confrontation with the flight experience becomes inevitable.

It is important to avoid labelling all those who experience fear as phobics. One important difference between phobic and non-phobic fear is said to be the involvement of avoidance in the phobic variety (Marks, 1969) the distinction seems rather to lie in the intensity of anxiety felt by so many of them in comparison with the milder anxiety felt by so few of the controls in response to similar environmental cues. Thus it might be more appropriate to think of a continuum of anxiety with the phobic subjects towards one extreme than to consider the phobics a discrete group.

Study 2

Study 2 consisted of a comparison of the effectiveness of seven treatment programmes designed to overcome fear of flying and a no treatment control group. Eighty-eight subjects from the original sample of 100 fearful flyers participated in this study. There were eight subjects in each group with the exception of the seventh treatment group which contained 32 subjects.

The first three programmes consisted of single intervention procedures. Treatment 1 involved group systematic desensitization modelled after Paul (1966); Treatment 2 was a variant of Treatment 1 and consisted of an audio-visual desensitization programme. Treatment 3 consisted of group exposure to the airport and a stationary aircraft, the radar section and control tower and also a flight simulator. Three other programmes involved a combination of these procedures. Treatment 4 involved systematic desensitization, audio-visual desensitization and exposure to the airport and a stationary aircraft. Treatment 5 combined all the components of Treatment 4 with the addition of exposure to the radar section and control tower, while Treatment 6 incorporated all the procedures of the first 3 programmes. Treatment 7 followed the procedures described in Treatment 4 with one important modification insofar as this programme was based at Dublin Airport with desensitization being conducted in a Boeing 737 aircraft cabin simulator. The eight treatment programme consisted of a no treatment control group.

On completion of intervention all subjects experienced exposure to the phobic situation by taking a 30 minute return flight on a domestic route.

All subjects were administered the Aeroanxiety Survey pre-treatment, pre-flight, post-flight and with the exception of the control group and Treatment 7, at a 6-month follow-up period. The Aeroanxiety Survey provided a basic measure to evaluate changes in aeroanxiety resulting from intervention. It also provided a measure of the following flight-related situations:-

1)   Flight preparation anxiety
2)   Pre-flight anxiety
3)   Crew directives anxiety

4) Take-off anxiety
5) In flight anxiety
6) Pre-landing anxiety
7) Landing anxiety
8) Total aeroanxiety

Results

All seven groups undergoing an intervention treatment exhibited a significant reduction with regard to their Total Aeroanxiety scores between baseline and post-flight testing. Treatment 6, exhibited the lowest mean score overall.

When the pre-flight testing is included, the Total Aeroanxiety scores continued to show highly significant decreases (p<.0001) in time, i.e. across baseline, pre-flight and post-flight conditions for all the intervention treatments. The control group showed no change.

Analysis of the Aeroanxiety subscales also showed that each intervention produced a statistically significant reduction in anxiety in each of the flight related situations. The greatest anxiety at time of baseline measurement was directed towards the flight itself, and it is noteworthy that the In-Flight anxiety subscale showed very substantial decreases in anxiety for all the treatment groups but not for the controls.

A follow-up study conducted six months later showed that reduction in Total Aeroanxiety had been maintained.

Discussion

One of the major findings of the present study was that, irrespective of type of intervention, number of components employed, duration of treatment or the environment in which each specific programme was conducted, all seven treatment groups demonstrated a significant reduction in Total Aeroanxiety, whilst no significant change was observed in the control group. Thus, there seems little doubt that the various treatments reduced aeroanxiety. Thus the reduction in anxiety seems a robust one which lasted at least six months, i.e. the duration of treatment plus follow-up.

In conclusion I would like to add that the most frequent measure of success in previous fear of flying studies has been the criterion of whether the subjects take a flight immediately following treatment. However, of more importance is whether subjects take any post-treatment flights. In order to evaluate treatment effectiveness and to establish how often subjects had flown as a result of intervention, a second follow-up survey was conducted by the author on behalf of Aer Lingus Irish Airlines one year later.

Results of this survey, in which 72 (87%) of 80 participants returned the follow-up questionnaires, revealed that 59 (81%) had taken 276 flights to destinations within the United States, Australia, the Orient and Europe. It is therefore postulated that these results provide further support for the long-term effectiveness of treatment for fear of flying.

References

Howard, W.A., Murphy, S.M., and Clarke, J.C. (1983). The nature and treatment of fear of flying: a controlled investigation. *Behaviour Therapy*, **14**, 557-567.

Marks, I.M. (1969). Fears and Phobias. New York: Academic Press.

Marks, I.M. (1978). *Living with Fear: Understanding and Coping with Anxiety*, McGraw Hill Book Company, New York.

Paul, G.L. (1966). *Insight vs Desensitization in Psychotherapy: An Experiment in Anxiety Reduction*. Stanford,California: Stanford University Press.

Walder, C.P., McCracken, J.S., Herbert, M., James, P.T. and Brewitt, N. (1987). Psychological Intervention in Civilian Flying Phobia: Evaluation and a three-year follow-up. *British Journal of Psychiatry*, 151, 494-498.

27 Description and psychometric evaluation of a self-report instrument for fear of flying assessment

Xavier Bornas and Miquel Tortella-Feliu
Department of Psychology, University of the Balearic Islands

Behavioural assessment and fear of flying

The attention that behavioural assessment and behaviour therapy has paid to fear of flying does not match the importance nor the social and economical repercussion of this problem -it has been estimated that 20-25% of general adult population suffer from significant anxiety in flying related situations and that 10% of them avoid to take a plane (Agras, Sylvester and Oliveau, 1969; Dean and Witaker, 1980, 1982)-

Personality inventories -as the MMPI, EPI or 16PF-, general anxiety scales -as the "Taylor Manifest Anxiety Scale" or the STAI-, general fear surveys -as the FSS-, psychophysiological records, have been the more frequently used procedures in the assessment of fear of flying. But, historically, there is a relative lack of specific, formal and well validated assessment instruments. In a review of the psychological literature on that topic we have found 12 different self-report instruments (Aitken, 1969,1972; Goorney, 1970; Solyom, Shugar, Bryntwick and Solyom, 1973; Girodo and Roehl, 1978; Traub, Grosslight and Boroto, 1982; Howard, Mattick and Clarke, 1982; Walder et al. 1982; Yaffe, 1987; Grusky and Reiss, 1987; Haug et al., 1987; Greco, 1989 and Doctor, McVarish and Boone, 1990). However, only few of them show some degree of formalization and their diffusion is very poor. An analysis of their psychometric properties (reliability and validity) is only available for three of these questionnaires: the "Questionnaire on attitudes toward flying" (Howard, Mattick and Clarke, 1982), the "Fear of Flying Scale" (Gursky and Reiss, 1987) and the "Fear of Flying Scale" by Haug et al. (1987); Johnsen and Hugdahl, 1990). Furthermore, the lack of formalization appears in more basic ways: in most cases the assessment instruments are just described, with little specificity, in the method section of fear of flying treatment reports. Only two publications are specifically devoted to the assessment as the

174

central topic (Gursky and Reiss, 1987; Johnsen and Hugdahl, 1990). The diffusion of these self-report instruments is, also, very poor: only four of them are used in other studies and the "Fear of Flying Scale" (Haug et al., 1987) is the that has been used more than two times. Whereas self-report measures of discomfort are often thought to be the more useful ones for the assessment of fear of flying (Doctor, McVarish and Boone, 1990; Haug, Brenne, Johnsen, Berntzen, Götestam and Hugdahl, 1987; Solyom, Shugar, Bryntwick and Solyom, 1973), it is paradoxical to find such a low diffusion and formalization of those self-report instruments. We fully agree with these researchers on the importance of the self-report measures. Both theoretical and clinical reasons support this point of view.

Firstly, this kind of measures fit the need of measuring the subject's responses (anxiety, physiological or cognitive arousal, discomfort...) to highly specific environmental contexts. It is true that other assessment strategies, as physiological records during exposure to flight-related stimuli , accomplish this requirement even to a large extend, but they are costly and they also have serious methodological problems.

Secondly, and perhaps more important, there is strong evidence (Doctor, McVarish and Boone, 1990; Haug, Brenne, Johnsen, Berntzen, Götestam and Hugdahl, 1987; Solyom, Shugar, Bryntwick and Solyom, 1973) supporting that self-reports are the more sensitive measures of behavioural change and the best predictor of subject's reactions in the assessed stimular condition.

The third reason is derived from practice. The use of several assessment strategies is quite habitual in the fear of flying literature. However, the analysis of results is primarily concerned with data derived from self-report instruments while other available data (e.g. physiological) are often ignored. This fact may be related not only with the sensivity and the predictive power of the measure but with other practice-based criteria that make its utilization more feasible: to fill up a questionnaire is an easy task and it has a very low cost (from both economic and time perspective). These advantages facilitate a more detailed monitoring of clients progress during treatment phases and follow-up.

By those reasons we have designed and tested, the "Fear of Flying Questionnaire" (Qüestionari de Por a Volar, QPV).

Description and psychometric properties of the "Fear of Flying Questionnaire" (Qüestionari de Por a Volar, QPV)

Description of the instrument: QPV structure

The QPV is a 34 items self-report instrument for assessing the degree of discomfort caused by series of situations related with flying. It is adressed to general population and the original version is written in Catalan language. Spanish translation is also available and an English version is enclosed below.

When assessing fear of flying it seems necessary to distinguish among three different kinds of fear evoking situations. Fear of flying does not only mean to be afraid to go into an airplane. Some previous situations, such as packing up or taking the boarding card, can evoke very strong fear responses in some people. Moreover those responses could be elicited by "vicarious" situations, for example, someone who has just arrived tells you that he or she has had a bumpy flight. Moreover, clinical experience and data from other investigations (Gursky and Reiss, 1987; Howard, Murphy and Clarke, 1983, Walder et al., 1987) point out that discomfort evoking situations are not the same for all the people, and then fear can be analyzed on the basis of fear evoking conditions. According to that, it may be interesting to include in the self-report instruments the different kinds of potentially fear evoking stimular conditions, in order to establish specific fear of flying patterns.To have information about those individual patterns would be of a great interest to fit treatment strategies to personal characteristics of the phobia.

By those reasons we have included three different kind of fear evoking situations in the questionnaire so that the QPV consists of three subscales: (a). The first one is made up of 17 items referred to situations taking place into the airplane (items 1,2,3,4,5,6,7,8,11,12,19,20,21,22,26,28,34) and we have called it "Flying subscale". (b). The second subscale, "Previous situations subscale" is made up of 11 potentially discomfort evoking situations that preceed the flight (items 9, 10, 13, 15, 16, 18, 24, 25, 27, 31, 33). (c). The "Vicarious subscales" is the third one and is made up of 6 items describing situations related with the transmission of information or the observation of flight related events (items 14,17,23,29,30,32). For each of the 34 items subjects are requested to rate the degree of discomfort that each situation induces to themselves on a 1 to 10 Likert type scale (1 no discomfort, 10 the highest discomfort). Thus the QPV provides four scores: one for each subscale and the total score which is the sum of the three subscales scores. Because of the number of items of each subscale is not the same, it is convenient to divide the subscale score into the number of items of this scale, obtaining a corrected score ranging from 1 to 10, which facilitates the comparison of the different scores.

Fear of Flying Questionnaire (Qüestionari de Por a Volar, QPV)

1. Half way through a flight I have the feeling that the plane reduce speed and then it accelerates again.
   2. The crew announce that in a few minutes we will be landing at the airport and so we have to fasten our seat belts.
   3. In mid flight I hear the announcement that we can unfasten our seat belt and seat belts light turning off.
   4. The plane has taken off and it is ascending.
   5. During the flight I hear a sound that I think is strange.
   6. The plane suddenly accelerates and I feel how it lifts up.
   7. The plane goes through a cloudy area.
   8. The plane descends gradually and approaches the runway.
   9. When I get up in the morning the day I have to fly and see that I will have to fly with bad weather.
   10. In the airport terminal building I go to check-in.
   11. During the flight it seems that the plane has a slight fall or passes through a hole or an air pocket.
   12. At a moment of the flight the plane shakes a lot.
   13. I am at home packing up for my journey by plane.
   14. I am at the arrivals terminal to meet some relatives or friends.
   15. I am going by car on the way to the airport to catch a flight.
   16. I am at home or at work and in a few minutes I will be going to the airport.
   17. I hear on the news that a plane has suffered a small accident without casualties in an airport of my country.

18. I find out that I have to fly to mainland or one of the islands (*specially referred to the Balearic Islands*).

19. I have fastened my seat belt and the plane starts to move along the runway (taxi out).

20. During landing operation I feel the wheels touching down the runway.

21. At landing, I feel the plane braking vigorously.

22. In the plane before take off, I am looking at air-hostesses' safety measures demonstration.

23. I hear on the radio or read in the newspaper of an aircraft crash with mortal casualties.

24. I know that I will soon have to go on a flight longer than three hours.

25. Sat in the departure lounge waiting for the door to be opened to board the aircraft.

26. Sat in my seat with the plane still waiting for take off.

27. I go up the stairs boarding the plane.

28. Everything seems calm but suddenly during the flight we are requested to fasten our seat belts.

29. A relative or a friend tells me that he has had a bumpy flight.

30. I see on TV the images of an aircraft crash.

31. When I go to sleep the night before flying.

32. I am driving or going by car along a road near the airport and I see a plane taking off.

33. I am in the departure lounge and a delay in my flight is announced few minutes before boarding.

34. During the flight I see a group of two or more people who have left their seat and they remain stand up at he end of the aisle while the airhostesses serve drinks or food with the serving-trolley.

Descriptive data and psychometric properties of the QPV.

In order to provide some initial data concerning descriptives and psychometric properties of the QPV, it was administered to a sample of 166 students of Psychology in the University of the Balearic Islands -44 males and 122 females with a mean age of 21.31 years-old and to a group of 8 people -all males with a mean age of 44 years-old- who were participating in a fear reduction training program at the British Caledonian Flight Training Center in Majorca. Fifteen days after the first assessment teh subjects of the students sample was retest. Only 66 form the 166 initial subjects filled up the questionnaire in the retest condition (53.01%). Moreover, people from the non-clinical sample were requested to answer the question "Do you have fear of flying?". Data from the persons who answered "yes" to that question were analyzed as conforming a subclinical sample (n=13).

In table 1, we reproduce the mean scores on the QPV for the three groups. No significative sex differences were observed, for that reason we present the data of all the subjects without distinguishing between males and females.

Reliability test-retest index. The test-retest correlation index of reliability, with a 15 days interval between observations, is r=.97 for the QPV as a whole and r=.96 for the flying situations subscale, r=.96 for the previous situations subscale and r=.92 for the vicarious subsacale.

Internal consistency. It was calculated with Cronbach's alpha procedure. Alpha indexes are .97 for the complete QPV and .95, .94 and .82 respectively for the flying situations, previous situations and vicarious subscales.

Table 1. Means and standard deviations on the QPV

	Non-phobic students sample (n=153)		Sub-clinical sample (n=13)		Clinical sample (n=8)	
	M	s.d.	M	s.d	M	s.d.
FLYING	2.49	1.06	6.51	1.68	7.22	1.14
PREVIOUS	1.63	.79	5.24	2.17	7.98	.91
VICARIOUS	2.79	1.19	5.71	1.62	7.08	.59
TOTAL QPV	2.26	.86	5.96	1.75	7.44	.59

FLYING: Flying situations subscale; PREVIOUS= Previous situations subscale; VICARIOUS= Vicarious subscale; TOTAL QPV=Questionnaire total score.

Table 2. Variance analysis of QPV's mean scores obtained by the samples

SCHEFFE F-TEST among samples

	F-TEST	Non-phobic vs. Subclinical	Non-phobic vs. Clinical	Subclinical vs. Clinical
FLYING	$F_{(2,171)}=137.56$**	77.81*	68.3*	1 (n.s)
PREVIOUS	$F_{(2,171)}=236.08$**	84.56*	165.35*	19.95*
VICARIOUS	$F_{(2,171)}=78.00$**	34.87*	47.92*	3.21*
TOTAL QPV	$F_{(2,171)}=191.96$**	90.78*	113.07*	6.04*

* $p< .01$ **$p=.0001$. FLYING: Flying situations subscale; PREVIOUS= Previous situations subscale; VICARIOUS= Vicarious subscale; TOTAL QPV=Questionnaire total score.

Validity and clinical usefulness. As we can see in table 2, the QPV is highly capable in discriminating subjects from the different samples (non-clinical, clinical and sub-clinical). Examining the scores obtained in the three subscales (Table 1), we can see that significant differences among groups are observed ($p<.0001$) in the three subscales but the previous situations subscale is the one with the highest differences. The special sensitivity of this subscale in discriminating among groups may be related to the fact that situations described in the other scales evoke some degree of unconditioned fear responses in all the subjects, including those without fear of flying, while only fearful people experience discomfort in situations occuring before the flight.These data support, in some way, the hypothesis that subscales based on the source of the fear are rellevant for the assessment of fear of flying. Moreover, the sensitivity of the QPV to clinical change, as it is observed in the assessment of subjects participating in a structured program for the control of fear of flying (Tortella-Feliu and Bornas, in a chapter of this book), could be taken as a sign of QPV's construct validity.

References

Agras, S., Sylvester, D. and Oliveau, D. (1969). 'The epidemiology of common fears and phobia'. *Comprehensive Psychiatry, 10*, 151-156.

Aitken, R.C.B. (1969). 'Prevalnece of worry in normal aircrew'. *British Journal of Medical Psychology, 42*, 283-286.

Aitken, R.C.B. (1972). 'A study of anxiety assessment in aircrew'. *British Journal of Social and Clinical Psychology,* II, 44-51

Dean, R.D. and Whitaker, K.M. (1980). *Fear of flying: impact on the U.S. air travel industry.* Boeing Company Document, BCS-00009-RO/OM.

Dean, R.D. and Whitaker, K.M. (1982). Fear of flying, impact on the US travel industry. *Proceedings Human Factor Society, 26th Annual Meeting,* (pp.470-473). Santa Mónica, CA: Human Factor Society.

Doctor, R.M., McVarish, C. and Boone, R.P. (1990). 'Long-term behavioral treatment effects for the fear of flying'. *Phobia Practice and Research Journal, 3*(1), 33-42.

Girodo, M. and Roehl, J. (1978). 'Cognitive preparation and coping self-talk: anxiety management during the stress of flying'. *Journal of Consulting and Clinical Psychology, 46*(5), 978-989.

Goorney, A.B. (1970). 'Treatment of aviation phobias by behaviour therapy'. *British Journal of Psychiatry, 117* (540), 535-544

Greco, T.S. (1989). 'A cognitive-behavioral approach to fear of flying: a practitioner's guide'. *Phobia Practice and Research Journal, 2*(1), 3-15.

Gursky, D.M. and Reiss, S. (1987). "Identifying danger and anxiety expectancies as components of common fears'. *Journal of Behavior Therapy and Experimental Psychiatry, 18*(4), 317-324.

Haug, T., Brenne, L., Johnsen, B.H., Berntzen, K.G., Götestam, K.G. and Hugdahl, K.(1987).'A three-system analysis of fear of flying: A comparison of a consonant vs a non-consonant treatment method'. *Behaviour, Research and Therapy, 25* (3), 187-194.

Howard, W.A., Mattick, and Clarke, J.C. (1982). 'The nature of fears of flying'. University of South Wales: unpublished manuscript.

Howard, W.A., Murphy, S.M. and Clarke, J.C. (1983). 'The nature and treatment of fear of flying: a controlled investigation'. Behavior Therapy, *14*, 557-567.

Johnsen, B.H. and Hugdahl, K. (1990). 'Fear questionnaires for simple phobias: Psychometric evaluations for a norwegian sample'. *Scandinavian Journal of Psychology, 31*, 42-48.

Solyom, L., Shugar, R., Bryntwick, S. and Solyom, C. (1973). 'Treatment of fear of flying'. *American Journal of Psychiatry, 130*(4), 423-427.

Traub, G.S., Grosslight, J.H. and Boroto, D.R. (1982). 'Locus of control in predicting differential response to a treatment for flight anxiety'. *Perceptual and Motor Skills, 55*,188-190.

Walder, C.P., McCraken, J.S., Herbert, M., James, P.T. and Brewitt, N. (1987). 'Psychological intervention in civilian flying phobia: evaluation and a three-year follow-up'. *British Journal of Psychiatry, 151*, 494-498.

Yaffe, M. (1987).*Taking the fear out of flying.* London: Newton Abbot.

28 Evaluation of a structured multicomponent program in the treatment of fear of flying

Miquel Tortella-Feliu and Xavier Bornas
Department of Psychology, University of the Balearic Islands

Introduction: fear of flying treatments

The analysis and treatment of fear of flying has been a topic of interest for psychology, and particularly for behaviour therapy, from II World War although the development of research and applications on this field has taken place from the sixties. There is a considerable amount of studies on the analysis and treatment of fear of flying. We found 55 articles indexed at Psychological Abstracts during the period 1967-1992. However, generally speaking, the investigations have a level of quality lower than that one could expect. Case studies as well as treatment applications carried out without control groups are very frequent among the revised literature. There are only seven controlled studies on psychological intervention procedures till 1992 (Beckham, Vrana, May, Gustafsson and Smith, 1990; Denholtz and Mann, 1974, 1975; Denholtz, Hall and Mann, 1978 -a follow-up of the former-; Girodo and Roehl, 1978; Haug, Brenne, Johnsen, Berntzen, Götestam and Hugdahl, 1987; Howard, Murphy and Clarke, 1983; Solyom, Shugar, Bryntwick and Solyom, 1973; Walder, McCraken, Herbert, James and Brewitt, 1987). A lot of behavioural strategies have been used in the treatment programs: systematic desensitization, exposure techniques, relaxation procedures, cognitive reestructuring, information on aeronautics and on the nature of anxiety-fear responses... The main conclusions obtained by controlled studies are: (a) behaviour therapy is always superior to other treatment strategies or to control conditions, (b) there are no significative differences on the efficacy of the several behaviour therapy techniques used in fear of flying treatment programs, (c) although no clear objective data are available, authors point out the central role of exposure.

An increasing trend to the development of structured multicomponent programs is currently observed. This is partly due to the fact that no single technique has been ound to be more effective than any other. The use of highly structured treatment packages is also very frequent when interventions are addressed to groups. Recently, several airline companies, aeronautical associations and private clinics i.e. Doctor, McVarish and Boone, 1990; Greco, 1989; Roberts, 1989) offer group treatment programs for the fear of flying to their clients. Although this kind of interventions have important advantages (low cost, different components of fear - physiological, cognitive,...- are specifically treated), some problems arise when an adequate evaluation of the participing subjects as well as of the program itself is omitted. Even though the patients usually report an improvement at the end of the intervention, and therefore programs are thought to be effective, the real efficacy of the program and that of each program component is far from clear. Usually, assessment is only carried out before and after treatment, thus obtaining just a measure of the effectiveness of the whole program. With this kind of evaluation, information about the effectiveness of each program component is not provided. In order to improve the multicomponent programs it seems necessary to determine which components are effective and which are not. Detailed assessment, of the participating subjects as well as of the program itself, is also needed to provide information adressed to match clients' fear patterns to treatment characteristics. In the present paper we describe the evaluation of a fear of flying treatment program.

Evaluation of a structured multicomponent program

The program developed by the Instruction Centre of the British Caledonian Company in Mallorca presents some of the characteristics of the multicomponent programs described above: structured evaluation of the participating subjects and of the intervention program itself was omitted, although it was apparently very effective (clients usually reported verbally a noticeable improvement at the end of the treatment). Then we were consulted for designing an assessment procedure in accordance with two objectives: (a) To determine the real effectiveness of the program and of its components. (b) To know if the relative effectiveness of each component depends on subjects characteristics. If this was true, the program could be adjusted to the individual fear responses pattern.

Structure of the British Caledonian treatment program

It was a five component treatment package including five 1,5 to 3 hour sessions, one session per day from Monday to Saturday. The intervention was entirely carried out in the British Caledonian Flight Training Centre in Mallorca and each program component was applied by a different therapist, belonging to the British Caledonian Company or contacted by the staff. We did not participate in the design nor in the application of the program. The components of the program were: (a). Aeronautical information. (2 hours) (b). Human arousal and anxiety responses information. (2 hours) (c). Rational emotive therapy. (3 hours) (d). Relaxation (3 hours) (e). Direct exposure (flying simulation experience). (1,5 hours)

Evaluation procedure

Eight adult male people required to participate in the British Caledonian treatment. They were randomly assigned to the experimental (n=4) or waiting-list control

181

(n=4) condition. The main instrument used in the evaluation of the participants wa the "Fear of flying questionnaire" (QPV, Qüestionari de Por a Volar), a 34-item self-report questionnaire designed by the authors.Subjects had to report in a 1 to 10 points scale the degree of discomfort that the 34 flying-related situations induced to themselves. The QPV is divided into three subscales: the "Flying situation subscale" (situations taking place into the airplane), "Previous situations subscale (potentially evoking situations tha preceed the flight) and the "Vicarious subscales (situations related with the transmission of information or observation of flight related events). A more detailed description of the QPV appears in another chapter of this book (Bornas and Tortella-Feliu). Psychophysiological measures were also taken but they are not included in our current analysis. Participating subjects were assessed five times along the treatment process: (a) before the treatment begin (initial assessment), (b) after information on aeronautics and on human arousal anxiety responses sessions, (c) after rational emotive therapy session, (d) after relaxation practice and (e) after flying simulation practice (final or post-treatment assessment). By this way not only the treatment package is assessed but also the relative effectiveness of each of the included components.

Results

The treatment package is quite effective in reducing fear of flying (see figure 1) The fear-discomfort total mean of the experimental group falls from 7.5 to 3.25 while control group scores, on the other hand, remain as high as before treatment Global discomfort degree at the end of the program is similar to that reported by non-clinical population, although the non-clinical sample discomfort degree is still lower than discomfort reported by experimental group after treatment. We think that this reduction has a high clinical significance.

In a more detailed analysis of the effects of the treatment, and attending to subscales scores (figure 2) we must pay attention to the fact that the more dramatic reduction is on the "flying situation subscale" (from 7.22 at base-line assessment to 2.75 at the end of the treatment), reaching a score very similar to the mean obtained by a non-clinical sample (2.49). On the other hand, the smallest reduction is on "previous situations subscale" (from 7.98 to 4.75). This is not consistent with the results of other researchs (i.e. Howard, Murphy and Clarke, 1983) where a multicomponent program including flooding, relaxation and systematic desensitization, was successful in reducing anticipatory fears. How can we explain this difference?.This effect may be due to the fact that the program was primarily adressed to flying situations and it did not take other fear evoking situations into account. It could be suggested to include some kind of training for coping with previous and vicarious fear-evoking situations in future applications of the program. Furthermore, it is interesting to realize that a salient characteristic of people who fear to fly is that they report the highest discomfort degree when they have to face situations that occured before the flight, such as packing up at home or waiting in the lounge for boarding. On the other hand, people without fear do not feel anxious in these situations although they usually report some low discomfort in front of stressing in-flight and vicarious conditions.

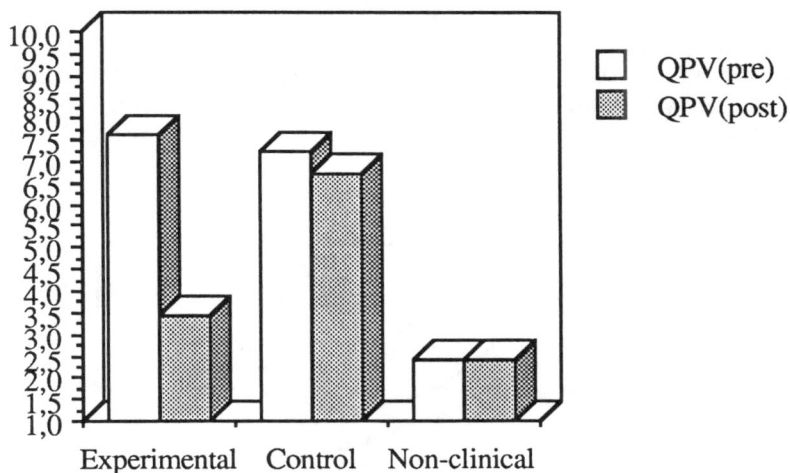

Figure 1   Total   scores   on   the   QPV   by   groups   before   and   after treatment

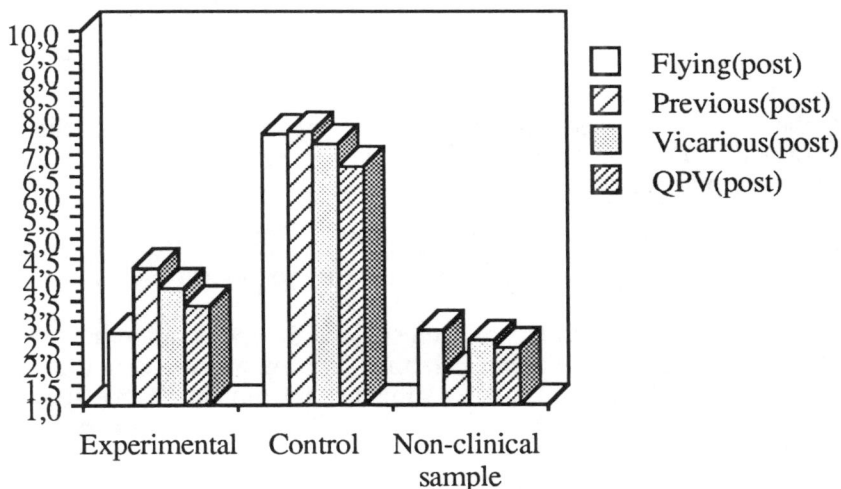

Figure 2   Mean   scores   on   each   QPV   subscale   by   groups   after treatment

Let us examine now the effectiveness of each component of the program. If we look at the mean score profile (see figure 3), we can see substantial differences. Apparently, the information and the direct exposure (flight simulator) components seem to be the real active ingredients of the program, while rationale emotive therapy and relaxation produce a very small fear reduction (and in some individuals

183

these components seem to increase their discomfort degree). Reductions after the application of the former components are evident for total score and for previous and flying subscales, not for vicarious subscale where the decrese is more gradual and less important. This is not new as the importance of direct exposure in the reduction of fear of flying, when analyzing multicomponent programs, has been pointed out by several authors (Howard, Murphy & Clarke, 1983; Solyom, Shugar, Bryntwick and Solyom, 1973; Walder, McCraken, Herbert, James and Brewitt, 1987)

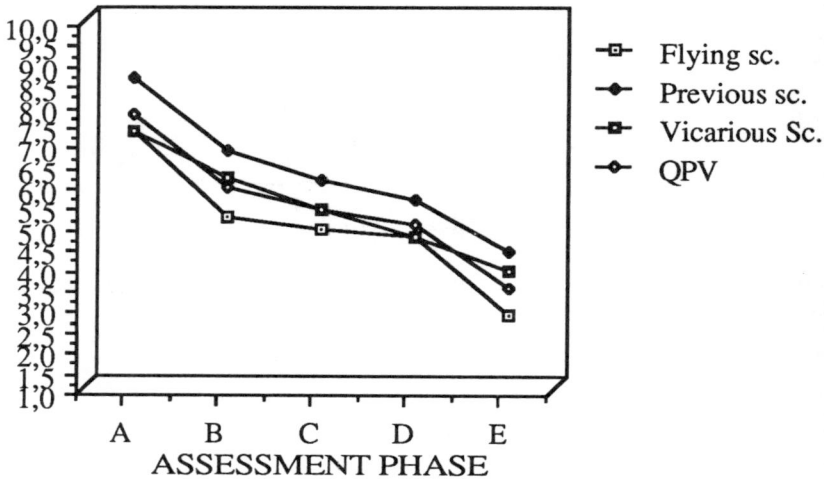

A=base-line; B=assessment after the two information components; C=assessment after RET; D=assessment after relaxation practice; E=assessment afetr flight simulator practice (exposure).

Figure 3 Mean scores obtained by experimental group after the completion of each component of the program

Looking at the score profile of each subject, the more salient factor is interpersonal variability. However we can establish the hypothesis that the efficacy of each program component in reducing subjects' discomfort degree seems to depend on QPV initial scores (base-line assessment). On one hand, information on aeronautics and on human anxiety responses program components have a more salient effect on subjects with a high global score at initial assessment. On the other hand, direct exposure (flight simulator), seems to be more effective with people who had moderate discomfort values on QPV at initial assessment. These data support the idea that tailoring the treatment to patient's pattern of reaction and/or to personal fear-evoking conditions would be a powerful strategy to be taken into account (i.e.Gursky and Reiss, 1987; Haug, Brenne, Johnsen, Berntzen, Götestam and Hughdahl, 1987).

All these results must be cautiously considered. More applications of the program are necessary for replicating results. Furthermore, the order of the components of the program would be changed to ensure that the effect of each component does not depend on the place it occupies into the treatment package.

References

Beckham, J.C., Vrana, S.R., May, J.G., Gustafson, D.J. and Smith, G.R. (1990). 'Emotional processing and fear measurement synchrony as indicators of tratment outcome in fear of flying'. *Journal of Behavior Therapy and Experimental Psychiatry,* **21**(3), 153-162.

Denholtz, M.S. and Mann, E.T. (1974). 'An audiovisual program for group desensitization'. *Journal of Behavior Therapy and Experimental Psychiatry,* **5,** 27-29.

Denholtz, M.S. and Mann, E.T. (1975). 'An automated audiovisual treatment of phobias administered by non-professional'. *Journal of Behavior Therapy and Experimental Psychiatry,* **6**(2), 111-115.

Denholtz, M.S.. Hall, L.A. and Mann, E. (1978). 'Automated treatment for flight phobia: a 3 1/2-year follow up'. *American Journal of Psychiatry,* **135**(11), 1340-1343

Doctor, R.M., McVarish, C. and Boone, R.P. (1990). 'Long-term behavioral treatment effects for the fear of flying'. *Phobia Practice and Research Journal,* **3**(1), 33-42.

Girodo, M. and Roehl, J. (1978). 'Cognitive preparation and coping self-talk: anxiety management during the stress of flying'. *Journal of Consulting and Clinical Psychology,* **46**(5), 978-989.

Haug, T., Brenne, L., Johnsen, B.H., Berntzen, K.G., Götestam, K.G. and Hugdahl, K.(1987).'A three-system analysis of fear of flying: A comparison of a consonant vs a non-consonant treatment method'. *Behaviour, Research and Therapy,* **25** (3), 187-194.

Howard, W.A., Murphy, S.M. and Clarke, J.C. (1983). 'The nature and treatment of fear of flying: a controlled investigation'. *Behavior Therapy,* **14,** 557-567.

Roberts, R.J. (1989).'Passenger fear of flying: behavioural treatment with extensive in-vivo exposure and group support'. *Aviation, Space and Environmental Medicine,* **60,** 342-348.

Solyom, L., Shugar, R., Bryntwick, S. and Solyom, C. (1973). 'Treatment of fear of flying'. *American Journal of Psychiatry,* **130**(4), 423-427.

Walder, C.P., McCraken, J.S., Herbert, M., James, P.T. and Brewitt, N. (1987). 'Psychological intervention in civilian flying phobia: evaluation and a three-year follow-up'. *British Journal of Psychiatry,* **151,** 494-498.

Part 7
HARDWARE AND SOFTWARE INTERFACE DESIGN

29 Systems engineering, cognition and complex systems

Iain S. MacLeod, Aerosystems International, Yeovil, Somerset, UK

Historical Development of Aircraft Instrumentation and Systems

Aircraft instrumentation and aircraft systems have developed from the early magnetic compass, pressure altimeter and airspeed indicator to current complex avionic aircraft systems. The early interactions between aircrew and their flight assisting instruments were always on a direct and simple one to one basis with the aircrew interfaced to the instrumentation /simple system through a single type of dial or control. In contrast, the Man-Machine Interfaces (MMI) of modern complex avionic systems have a tendency to over complicate the aviator's view of the flight and its environment.

Advances in computing have promoted advances in avionics and also an associated burgeoning avionic system complexity. This emphasis on technological complexity in the design of Man-Machine Systems has promoted automation of many functions previously left under the control of the human.

Poorly considered automation may unintentionally relegated the human from a flight / aircraft operation control role to that of a poorly informed supervisor or observer of fragments of the functions, states and processes of aircraft avionic systems. A ill informed supervisor is badly placed to take control in a emergency. Further, the interfaces between the aircrew and the automated and complex avionic systems monitoring and flight control are frequently multi functional in design. Therefore, they can display inappropriate and untimely forms of information and cues to support the requirements for real time human supervision and control.

The relatively recent drive for automation in aviation has been partly funded by promises of cost savings through reductions in crew numbers. Such reductions leave the reduced but already trained aircrew with the same individual and collective flying skills. However, in addition they have the difficult task of reliably supervising more functionality and processes than before, with less control and with the safety of the aircraft still their prime responsibility.

System Life-Cycle Considerations

The above consideration on historical development suggests consideration on aircraft and their systems with relation to their origins and life-cycle. The definitions of a system life-cycle are many, emanating from the multi disciplinary standpoints involved in system engineering. An attempt at a full definition will not be made here. Suffice to say that a system life cycle starts with a concept, is developed through various stages of design, undergoes some form of evaluation, if found acceptable is then adopted and serves towards an intended purpose, undergoes improvements during its working life and eventually is made redundant.

Until very recently the main emphasis on system design and development was on engineering aspects and an associated use of the most up to date technology commensurate with acceptable design risk. The design risk was not seen as residing in any lack of due consideration on the future needs and capabilities of the aircrew within the new system.

New technology is offered with arguably a greater capability and reliability than that available from previous systems. A complex avionic system is designed to efficiently achieve more in order to expedite the equation of system costs, and thus to increase the life-cycle profitability of ownership. However, complex systems are hard to understand. Further, it must be questioned whether the necessary safety margins in the operating environment are maintained or decreased in the light of the greater engineered reliability and efficiency of the new system (see Hollnagel 1993 on risk homeostasis).

Evidence suggests that safety margins can be adversely affected as improved performance frequently results in a belief that the margins can be decreased. Such a narrowing of safety margins also fails to appreciate the often uncertain contribution of the human component within the engineered complex system - goal directed and skill mediated human cognitive functions, as applicable to the maintenance and operation of the system, having been ignored throughout design.

A tenet of this paper is that only if human cognition and mental strategies are considered and appreciated as important system elements throughout the system life cycle, can the design of complex avionic systems hope to be certified as reliably 'fit for purpose' (Taylor and MacLeod 1993).

Life-Cycle Application of Traditional Human Factors (HF) Methods

Aerosystems International's (AeI) involvement in several aircraft programmes suggests that an application of traditional Human Factors (HF) methods to complex man-machine systems' life-cycle design may suffer from the following shortfalls:
- An over strong emphasis on technology and the natural sciences;
- A foundation on laboratory findings rather than on 'real world' practices;
- The application of HF to only part of the system life-cycle with little resulting HF traceability throughout the system design process;
- A failure to consider human goals with relation to human aspirations, human limitations in decision and performance, cognitive skills and capabilities;
- Human cognitive complexity, perhaps because it is difficult to bound, is seldom considered thus promoting uncertainties on system performance;
- A concern more with maintaining the acceptability of HF standards by the engineering community, and these standard's ability to proffer general advice, than with their level of their true practical applicability to complex systems design;
- Traditional HF tools are poorly founded on theory because many of the existing related theories are frequently found to be wanting in practice;
- The other system engineering disciplines often tend to decry 'Professional' HF assistance and instead adopt 'Common Sense' as their practical tool;
- Currently, HF cannot reliably deal with complex man-machine systems.

Generic Problems Arising through Complexity

A simple answer to a complex problem is likely to be a wrong answer. Complex systems may breed uncertainty and the possibilities of many uncontemplated errors in system performance. A poorly designed man-machine complex system results in the human component of the system struggling to comprehend the remainder of the system, rather then working through that system within an environment in order to effectively achieve system goals. As a result the performance of complex man-machine systems can be degraded by untimely and unnecessary human interventions.

One of the frequent problems in the design of real time man-machine complex systems is that a designed and engineered system micro functionality is presented unrefined to the human at the MMI. This can lead to conscious or subconscious 'mindless' behaviour (Langer 1992) where the human perception is swamped with inappropriate micro representations of system functionality. Needed information is frequently missing or poorly represented; for example, information on system states, the efficiency of processes, performance being achieved in the environment, the trends in actual performance and the criticality of these trends on system goals.

An Approach to Complex System Development

Over the last few years, AeI has attempted to form a synergy between the engineering approach to functional based specification and design of aircraft

191

and a pragmatic approach to addressing the cognitive needs of aircrew within system design. This approach has been applied to recent aircraft development projects with some success. The approach has the following tenets:

- That human understanding within the system must be promoted;
- That the limitations and capacities of human cognitive processes must be considered within the system design process;
- That unwanted workload must be ameliorated where possible;
- That the limits of human attention, especially selective attention, must be considered;
- That HF tools should be developed with the intention that they be applicable throughout the system life-cycle (MacLeod et al 1993)
- That information must be available to the human appropriate to their needs for that information (e.g. considering what, where, when, form, rates, cues, feedback, advice, cautions etc);
- That the system design must aid the human to maintain situational awareness, support effective decision making and promote human reliability at work.

The following is a brief summary of the method:

Conceptual Development Phase

At the conceptual design phase the systems engineer must consider the requirements of the system and the high level engineering functionality that meets these requirements. Earlier but similar systems must be considered in that previous systems should be a source of advice on what was good, and worth retaining, and what was bad and should be discarded.

The purposes of system functions, the associated information flows, system activities, control nodes and products within the system should be broadly determined and grouped (system processes). Thus a high level function will encompass processes considering sub functions and their processes.

High level flight performance profiles can then be constructed and the processes within time segments of the profile considered. Use of Subject Matter Experts (SMEs) should be made at this stage. SMEs are especially useful in helping to determine the most critical segments of flight with relation to human performance. Cognitive task analysis techniques, including questionnaires, are used with SMEs to assist the selection of these segments. Post the segment selection, and through use of collective knowledge of appropriate system processes, 'story-lines' are composed to present the best envisaged story for the critical human and system performance of each segment. These story-lines are revised as necessary.

The story lines are used as a basis for iteratively examining any conceived human roles and jobs within the system. Using cognitive task analysis methods appropriate to available expert knowledge and appreciating existing like systems, the critical tasks for the segments are considered. The story-lines are then revised and amended as necessary. The most critical tasks are then further examined for seemingly inappropriate loads on attention, memory or time pressure to complete activities.

The determined story-line tasks are modelled (the sophistication of the modelling related to the level and quality of data available) and the models discussed with the SMEs and revised as necessary. The tasks are then subjected to a more detailed and predictive version of cognitive task analysis to determine what information and conditions appear necessary for the performance of tasks, are needed to support task related plans / alternative plans and ameliorate work 'bottlenecks'.

Post the performance of the above analysis, SMEs are again consulted as appropriate to determine, in the light of their experience, the information and cues that they consider are necessary to complete the considered tasks. As an example, cues may be categorised as to their appropriateness or inappropriateness for user requirements depending on tasks and system status.

The above findings are compared to the previous information deemed necessary by the detailed cognitive task analysis; to the overall groupings of system processes; to the initial conception of roles and jobs - all to arrive at early estimates of the number of people, workstations etc. required to achieve an equitable initial solution to the complement of man and machine within system design.

System Design Phases

The system design phase uses iterative design processes to continually update already obtained concepts, human engineering data and models. Here the early formulation and agreement of a Human Engineering Programme Plan (HEPP) and Test and Evaluation Plans (T&E) is important as a means of determining the application, purpose and expected outcomes of Human Factors application throughout the design process. Especially important are these plans' integration into the overall system management and development plans.

An early use of mock-ups and prototypes should be used to help to update the earlier concepts. Care should be taken to ensure that mock-ups and prototypes are not used solely for demonstration but are included in the evaluation programme. AeI uses task analytic simulation at this stage, to model and examine aircraft system operations (MacLeod et al 1992). These simulations assist in the stochastically prediction of workload, the discovery of system error provoking activities and the detection of attention bottlenecks. The simulations also act as a focus for SMEs and design engineers to discuss potential problems in system operation. Here, and at other earlier and later stages, consideration on the tenets of Naturalistic Decision Making has good practical value (Klein et al 1989). At this juncture, major problems in HF aspects of design can often be detected and are relatively easy to rectify.

Less major problems can also be considered, and through trade-offs with other system engineering areas, solutions can be found within the costs of the project. These solutions may result in early design changes, a change in emphasis with on-going design, changes to recommended procedures or advice on the future training emphasis.

As design progresses through the life-cycle, the choices made in the past must be checked and modified as necessary. Under the T&E Plan and HEPP, a progressive acceptance of the design should be attempted with the customer.

Above all, a traceability of the HF methods, associated design decisions and tests should follow throughout the design process to ensure the efficacy of their application within the progression of the design.

Pointers

The above coverage is naturally much simplified. Important pointers to the methodology are:
- An acceptance that some of the most important functions in a man-machine system may be cognitive;
- A concern with real world situations under an appreciation of the evidence available at each stage of the system life-cycle;
- A focusing onto critical functions and tasks. A refinement or detailing where necessary to determine the nature of the criticality;
- Close continual involvement with the customer plus SME assistance;
- An adoption of cognitive task analysis methods;
- The striving for synergy of design through regular liaison and trade-offs with other disciplines;
- The belief in the efficacy of a good and early agreed story-line, subsequent related predictive modelling, model refinement and early testing against real world evidence.
- The appropriate use of modelling techniques, mock-ups, prototypes and simulations to support a progressive acceptance of the system under the auspices of the HEPP and T&E Plans.
- A belief that for the sake of economy the light at the end of the tunnel will not be switched off.

References

Hollnagel, E. (1993) Human Reliability Analysis: Context and Control, Academic Press, London.

Klein, G.A, Calderwood, R and MacGregor D (1989) Critical Decision Method for Elicitating Knowledge, *IEEE Transactions on Systems, Man, and Cybernetics,* **Vol. 19**, (No. 3), May/June.

Langer, E.J. (1992) Matters of Mind: Mindfulness/Mindlessness in Perspective, *Consciousness and Cognition*, **1**, pps 289-305

MacLeod, I.S., Biggin, K, Romans, J and Kirby. K (1992) Predictive Workload Analysis - RN EH101 Helicopter, *Contemporary Ergonomics 1992*, Taylor and Francis, London.

MacLeod, I.S., Farkin, B and Helyer, P (1994) The Cognitive Activity Analysis Toolset (CAATS), *Contemporary Ergonomics 1994*, Taylor and Francis, London (In Press).

Taylor, R M & MacLeod, I S (1993) Quality Assurance and Risk Management: Perspectives on Human Factors Certification of Advanced Aviation Systems, *Proceedings of the Workshop on Human Factors Certification of Advanced Aviation Technologies, Toulouse, France*, 19-23 July (In Press).

30 A review of the benefits of colour coding collimated cockpit displays

Helen Dudfield, Defence Research Agency, Farnborough, UK

Introduction

The development of novel technologies, such as the Penetron CRT (Banbury, 1990) which has the ability to produce a limited colour set, has opened the debate on the utility of colour Head-Out Displays (HODs) such as Head-Up Displays (HUDs) and Helmet Mounted Displays (HMDs). The objective of this research is to identify whether colour coding is beneficial on such surfaces.

Operators tend to prefer colour, finding it a natural cue. In the psychological literature there have been numerous papers on the principles of colour coding and its benefits (Christ, 1975; Teichner, 1979). Traditionally, colour can be used on cockpit displays (Reising and Aretz, 1987) to encode busy tactical situation displays on which a pilot has to search for targets, or on system status displays to highlight information. Moreover, colour coding may stimulate preattentive processing which requires no conscious attention (Previc, 1988). In contrast, colour will not help when it is used for unimportant information, to classify many types of data, or to cue small or peripheral symbology.

The advisability of colour coding the HUD has been questioned. A recent HUD review advised that "Colour should not be used for HUD symbology" (Weintraub and Ensing, 1992, p.52). It is essential then that any incorporation of a multichromatic HOD into a future cockpit is dependent on evidence that the symbology can be adequately perceived and that such a HOD has clear advantages operationally over monochromatic HODs.

The utility of colour will depend not only upon performance benefits but also on ergonomic aspects, such as colour contrast and readability (Silverstein, 1987), and proof of technological performance. The HOD operates in a harsh environment. It has to be able to operate in high ambient lighting conditions, to cope with the effects of rapid manoeuvring and 'g', and to be extremely reliable. Finally, any proposed system would have to be compatible with image intensifiers such as Night Vision Goggles (NVGs). Thus, it is necessary to take into account physiological, psychological, environmental and technological factors. For example, research is necessary concerning both the physical performance of technologies and the perceptibility of colour on a collimated glass display against a multi-chromatic environment.

Previous colour HOD research

To date, there has been little research into the advantage of colour coding HODs. In one example, a HUD contained a pathway in the sky that was colour coded (Hawkins, Reising and Gilmour, 1983). The results indicated that errors in route following were reduced by 25% with the coloured pathway in the sky compared to the monochrome version, and pilots rated the colour HUD higher than the monochrome version on a useability scale. A further use of colour coding on the HUD that has been investigated to a limited extent is its application to cueing attitude information. A survey of 56 F-18 pilots found that 39 pilots responded positively to the addition of colour for attitude discrimination (Roust, 1989). This was experimentally evaluated by Zenyuh, Reising and McClain (1987) who compared colour pitch bars (blue for sky, green for ground, white for the horizon bar) with a monochrome format, using recovery from unusual attitudes as a measure of performance. In addition to colour, articulated bars were compared to parallel bars. The data analysis found a trend towards better performance in extreme conditions with coloured articulated bars than with coloured parallel bars. Typically, though, while subjects prefer colour coding, performance differences are difficult to prove (Martin and Way, 1987; Reising and Aretz, 1983). A recent survey carried out by Baird (1993) found that 30% of a sample of 69 RAF aircrew wanted colour added to ten HUD items. Many were related to limits of a systems: speed limits, a 'g' indication and target in range. Others were specific warnings: radar lock, low height, engine fire, missile launch, and fuel and defensive aids warnings.

A review of colour coding experimentation

The experiments that are presented below represent a summary of a considerable effort to investigate the benefits of colour coding HUDs.

Experiment 1: Use of colour for flight path awareness

In this first experiment, the philosophy used to colour code a HUD format was based on economy, i.e. colour was used only when the performance of subjects was outside pre-determined boundaries (e.g. flying off course), providing subjects with immediate feedback on the accuracy of their performance (Dudfield, 1992).

Performance was compared between a conventional monochrome, green HUD format and two redundantly colour coded alternatives, a Penetron HUD (three colours) and a full colour HUD (five colours), in a 'pilot-in-the-loop' simulation. The participants were either experienced at flying simulators/civil aircraft or novices. For inexperienced subjects, the availability of additional redundant information might facilitate the learning process but for the experienced group such information might not assist unless their workload was increased either by increasing display clutter or task complexity.

The type of HUD (colour or monochrome) and experience of the subject (novice or expert) were manipulated, and rms (root mean square) error data on deviation from a required height, speed, course and attitude were recorded. The task of subjects was to maintain a certain flight profile using simulations of three different HUDs. The task items were the airspeed indicator (ASI), heading scale and altimeter. The colour of all elements of the HUD was green, but in both colour conditions, redundant colour coding (amber and red) indicated flight path inaccuracy under the following conditions: if a subject veered off course, flew too low or high, too fast or slow or with a high degree of roll, the relevant HUD item changed from green to amber and ultimately to red, depending upon the amount of deviance from a set of flight profile criteria.

Colour was found to reduce workload, as measured by the NASA TLX. In particular, the colour Penetron simulation was reported to reduce the temporal demand and the amount of frustration that subjects experienced. However, the objective data failed to demonstrate performance benefits of colour.

Experiment 2: Use of colour for flight path awareness II

A second experiment was carried out with the same task as before, but with RAF aircrew, as well as simulator experienced personnel, as subjects (Dudfield and Hughes, 1993). As before, there was no evidence of a reduction in rms error with colour. However, in this case, there was no subjective reduction in perceived workload, although comments by the aircrew were favourable of colour coding on HUDs. Possibly, since the subjects were all experienced in flight or simulated flight, the task was not sufficiently demanding to produce performance benefits. In addition to rms measures, this trial assessed performance by eye movement measurement. Since the primary source of information was visual, it was anticipated that, by recording the spatial location and duration of fixations, information gathering strategies could be identified (if such strategies existed).

Overall, there were few indications in fixations or dwell time that colour reduced the time to perceive HUD information. There was a single exception: dwell times with the colour coded ASI were shorter than with the monochrome version by an average 20ms. Hence, the colour code on the speed indicator was, to a limited extent, assisting in speed maintenance. This difference was small, but in terms of the number of fixations made during typical HUD operations might have practical advantages. There was also an indication that the subject experience and HUD type interacted for height performance, since aircrew had shorter fixation durations with the colour HUD than the monochrome but the reverse was true for non-aircrew.

Experiment 3: Use of colour for attitude awareness

One of the main criticisms of the first two colour experiments was that colour was used to code the accuracy of a subject's maintenance of a flight profile but was not used for realistic HUD tasks. For example, the height read-out was coded such that it changed colour if subjects were below or above the required height. Yet as Baird (1993) found, aircrew wanted colour to provide a ground proximity warning, such as below 150ft. Hence, colour would seem to be more appropriate when applied uni-directionally.

An example of an operational use of colour on the HUD was provided by experiment 3 (Dudfield, in press). To test further the hypothetical advantage of the colour HUD, this experiment concentrated on a high workload task which aircrew hope never to experience operationally: recovery from an unusual position (UP). Such an event can be due to loss of consciousness during high g manoeuvring, spatial disorientation or pilot error. If the pilot realises that he is in a UP, he has to go through a set recovery procedure: a roll and pull to the nearest horizon. In this research, the accuracy and speed with which this recovery was achieved were measured when the HUD pitch bars were displayed in monochrome or colour. As with Zenyuh et al (1987), the sky/ground distinction was exaggerated through the use of coloured pitch bars (blue positive pitch bars and green negative pitch bars). Subjects were presented with a series of UPs and the initial reaction time (RT) to first stick input was recorded, as was the time taken to complete the recovery.

There were no significant differences between the monochrome and colour HUD conditions. Nonetheless, there were indications in the data that the colour pitch bars improved performance in certain cases. In the initial reaction time to UPs, there was a trend suggestive of an advantage for aircrew with the coloured pitch bars. It appeared that aircrew responded faster with the blue/green pitch bars than with the monochrome bars. The reverse was true for novice subjects. This finding was complemented by trends in the total time to recover to the horizon: response times were reduced with the coloured pitch bars as compared to the monochrome bars. Further, aircrew benefited more from the former, recovering more rapidly with the blue/green combination than with the conventional format. The majority of subjects commented that the blue/green difference was an effective cue for recovery strategies.

Conclusions

In this series of trials, the benefits of colour were difficult to demonstrate objectively. However, as Reising and Aretz (1983) found there was always a strong subjective preference for colour coding. This desire for colour has been found in many applied trials where objective support is lacking. It may be that colour benefits are not being sensitively measured or that the processing differences are too insignificant to be detected. The exceptions to this seem to lie in the development of ecologically valid tasks (such as recovery from UPs) and measures (e.g. eye movement recording). The lack of support for the incorporation of colour in the cockpit is a common research finding. The conclusion of research by Reising and Aretz (1983) was that in the absence of adequate objective metrics, "it may be that the researcher will have to support the use of colour on less scientific but possibly more powerful grounds - the users like and want colour" (Reising and Aretz, 1983, p.7). Burnette (1984) concurs with the use of colour on aesthetic grounds which are regarded as "those situations where pilots desire colour but no demonstrated performance benefits have as yet been shown to exist" (Burnette, 1984, p.1356).

From surveys of aircrew, it is obvious that colour needs to be applied to specific flight tasks (Roust, 1989; Baird, 1993). The use of colour to code pitch bars illustrates this. Recent research by Hardiman, Dudfield and Lal (in press) has confirmed that colour coding pitch bars assists a subject to recognise the attitude of his aircraft. A combination of colour (blue/green) and pitch bar shape (articulated/tapered) asymmetry enabled subjects to reach a response decision more rapidly than with monochrome symmetrical pitch bars.

Other possible examples of colour coding on a HUD are specific warnings regarding threats and the transgression of system limits. Baird et al (1993) considered using colour to draw pilots' attention to specific events. Ten HUD items selected from Baird's (1993) survey were displayed on a HUD format either in monochrome or colour. The results showed that colour coding was beneficial when applied to a digital read-out that had exceeded a critical limit. RTs for speed limits and low height warnings were significantly longer, and accuracy was worse, when the warnings were presented in monochrome than when they were presented in colour.

In addition to using colour to cue informational events (Teichner, 1979), colour may also have a role as a source of declutter, distinguishing between different types of HOD information (e.g. navigation and weapon aiming data). Finally, colour is being applied to the concept of a virtual cockpit where the pilot may be fully immersed in an artificial representation of the real world.

Acknowledgements

I wish to acknowledge the efforts of my co-researchers, in particular, Dr P Hughes (DSTO, Australia).

References

Baird, J-A. (1993), *A Survey Of Pilots' Attitudes To Colour Coding Head-Up Displays*, Defence Research Agency Technical Report 93030, Farnborough, UK.
Baird, J-A., Dudfield, H.J., Davy, E. and Moore, F. (1993), *A Study To Investigate The Benefits Of Colour Coding Based On The Results Of An Aircrew Survey.* Defence Research Agency Technical Report DRA/AS/FS/TR 93014/1, Farnborough, UK.
Banbury, J. (1990), 'Colour in Head-Up Displays', Paper presented at the *U K*

Chapter of the Society for Information Display Symposium, Basildon, UK.

Burnette, K.T. (1984), 'Multi-colour display design criteria', in *Proceedings of the National Aerospace and Electronics Conference*, Dayton, Ohio, May 21-25, 1348-1363.

Christ, R.E. (1975), 'Review and analysis of colour coding research for visual displays', *Human Factors*, **1** 7(6), 542-570.

Dudfield, H. J. (1992), *An Experimental Evaluation Of A Colour HUD*. Defence Research Agency Technical Memorandum FS 1033, Farnborough, UK.

Dudfield, H.J. (in press), *The Use Of Redundant Colour Coding For Attitude Awareness*. Defence Research Agency Technical Report, Farnborough, UK.

Dudfield, H.J. and Hughes, P.K. (1993), *Colour Coding As An Alerting Mechanism In A Fast-Jet Head-Up Display*, DRA Technical Report DRA/AS/TR 93018/1, Farnborough, UK.

Hawkins, J.S., Reising, J.M. and Gilmour, J.D. (1983), 'Information interpretation through pictorial format', in *Proceedings of the Aerospace Behavioral Engineering Technology Conference*, Society of Automotive Engineers, 243-248.

Hardiman, T.D., Dudfield, H.J. and Lal, R. (in press), *An Experiment Investigating The Use Of Asymmetrical Pitch Bars In Head-Up Displays*. Defence Research Agency Technical Report, Farnborough, UK.

Martin, R.L. and Way, T.C. (1987), 'Pictorial format displays for two seat fighter aircraft', in *Proceedings of the 31st Annual Meeting of the Human Factors Society*, Human Factors Press, Santa Monica, USA, 1072-1076.

Previc, F.H. (1988), *Towards A Physiologically Based Symbology*. USAF School of Aerospace Medicine, Brooks Airforce Base, Texas, USAFSAM-TR-88-25.

Reising, J.M. and Aretz, A.J. (1983), 'Colour coding in fighter cockpits - it isn't black and white', in *Proceedings of the 2nd Symposium of Aviation Psychology*, Columbus, Ohio.

Reising, J.M. and Aretz, A.J. (1987), 'Colour computer graphics in military cockpits', in H.J. Durrett (ed), *Colour and the Computer*, Academic Press.

Roust, L. (1989), *Evaluation Of HUD Display Formats For The F-18 Hornet*, Naval Postgraduate School Thesis, AD-A208 651, Monterey, California.

Silverstein, L.D. (1987), 'Human Factors for Colour Display Systems: Concepts, Methods and Research', in H.J. Durrett (ed),*Colour and the Computer*, Academic Press.

Teichner, W.T. (1979), 'Colour and visual information coding', in *Proceedings of the Society for Information Display*, First Quarter, **2** 0(1).

Weintraub, D.J. and Ensing, M (1992), *Human Factors Issues in Head-Up Display Design*, CSERIAC State-of-the-art report.

Zenyuh, J.P., Reising, J.M. and McClain, J.E. (1987), 'Advanced head-up display symbology: Aiding unusual attitude recovery', in *Proceedings of the 31st Annual Meeting of the Human Factors Society*, Human Factors Press, Santa Monica, USA,1067-1071.

31 Interaction with intelligent cockpit systems: an analysis

Adrian W. Coxell, MMI Dept, Defence Research Agency, UK

Introduction

Operators experience difficulty in the monitoring and control of complex systems used in various aerospace applications such as cockpit or air traffic control systems. A striking example is the Hubble Space Telescope which has 6000 sensors that must be monitored in real-time (Laffey, 1991). Such operator tasks could be supported by artificial intelligence (AI). However, it has been reported that only 10% of all medium- and large-sized expert systems are truly successful (Keyes, 1989). For example, the failure of an AI application for an air traffic control system used by the United States Navy has been documented by Sloane (1991). Such poor performance may be due to factors intrinsic to the system itself, such as poor knowledge acquisition or inappropriate knowledge representation (Madni, 1988), or to sub-optimal interaction between the system and the user. Card (1988) has identified three aspects of the interaction between a human and an intelligent system (IS): the role of the IS; communication between the user and the IS; and automation within the system. This chapter considers research pertinent to each of these aspects and also considers the interface implications of real-time system control.

Role of the intelligent system

The role of the IS has profound implications for the amount of control and involvement that a user has in the decision-making process, and the tasks that the user must perform. There are several potential roles:

Intelligent system as adviser

When the IS acts as an adviser, the user is required to supply it with data and then to accept or reject its advice (Woods, 1986). These roles assigned to the IS and the user produce several problems. For example, the underlying assumption that the IS is more expert than the human has been found to be incorrect in many aerospace applications (Malin et al, 1991); and the allocation of control of the interaction to the IS is likely to degrade the performance of the human-machine system (Hoogovens Report, 1976).

Intelligent system as subordinate

Here, the IS functions as a semi-autonomous system capable of accessing data and controlling a particular process without intervention from the user. In such situations, the user acts as a supervisory controller. It is necessary that the user can coordinate activity with the IS if their areas of responsibility overlap, and the IS must be able to inform the user of system errors in a safe and reliable manner (Malin et al, 1991). Unfortunately, there is evidence that humans are poor supervisory controllers (Billings, 1991).

Both the advisory and the subordinate IS roles rely on the user being able to judge the appropriateness of information provided by the IS, but this ability has not yet been demonstrated (Woods, 1986). Further, there is evidence to suggest that where humans merely supervise an automatic controller they do not develop the skills necessary to handle faults in the absence of machine support (Kessel & Wickens, 1982). Furthermore, users in such systems may find themselves in a 'responsibility/authority double-bind' (Woods, 1986), always over-riding the IS if their trust in it is low, or always accepting the IS's decision if the cost of over-riding it is too high. Finally, operators may be totally discouraged from using a system that involves the user in inappropriate or minimal interaction since they may "fear that massive system failure may suddenly transfer ...total control...to the human decision process and that such an event would place...personnel in a situation that they could not handle" (Sloane, 1991).

Intelligent system as cognitive tool

The problems discussed above have led to the suggestion that the IS should support rather than supplant the role of the user in decision making (Malin et al, 1991; Woods, 1986). In this role as a cognitive tool, the IS is designed not to produce an 'answer' but to provide better information display and system management and to reduce the amount of information that the human must process. A further advantage is that the user does not have to monitor the advice from the IS in addition to all the other available data. Obviously, this approach ensures that the user remains in the decision-making/control loop and is responsible for all actions. However, it is probably not possible for an IS to function solely as a cognitive tool (Malin et al, 1991), since many applications will require some degree of autonomous IS operation due to system complexity and/or real-time constraint.

Communication

Safety-critical real-time systems require that operators receive accurate, appropriate and timely information. However, the information communicated may be distorted by various interface elements. For example, message lists do not adequately represent the time course of relationships between events, and schematic representations do not adequately convey states unrelated to the structure of the system, such as system modes (Malin et al, 1991). Advisory systems present problems concerning the advice and explanations provided. It has been suggested (Woods & Roth, 1988) that advice should be presented in a graduated manner, becoming more prescriptive and inflexible with increasing severity of the situation. However, it is clear that such an approach is valid only if the IS 'knows' the correct course of action. Explanations of system reasoning create further difficulties, since the user must know when to seek an explanation, and since explanations in many real-time situations are post-hoc justifications of prior actions or decisions made by the IS (Wick & Thompson, 1989).

Although it may be preferable to use an IS as a tool rather than as an advisory system, communication between the user and the system may be more complex. Advisory systems can function adequately by providing explicit instructions and explanations justifying their decisions, but cognitive tools may need to provide more information to the user. Hollnagel (1986) has suggested that this problem should be solved by 'restructuring' the problem situation such that the information provided to the user capitalises on the human strengths of pattern recognition, detection of novel situations (and similarities), and generalisation. This restructuring should also reduce the need for information transformation and interpretation by the user and provide information at the most appropriate time to reduce the load on memory and attention. Clearly, it is difficult to design a system interface that is able to indicate all of the important variables and system states, and the interactions and dependencies between them, without clutter and in such a way that the user is able to make timely decisions. An explicit understanding is required of the cognitive demands of the tasks that an IS is designed to support. For example, diagnosis is difficult because of the high ratio of values that must be kept in working memory to the amount of values that are visible (Wickens, 1992); any interface that improves this ratio should facilitate decision making.

Automation

Aircrew experience considerable difficulty in the monitoring and control of conventional automated systems (Billings, 1991). Currently these systems contain three types of automation (Fadden, 1990). Control automation is used either to assist or replace the pilot in the task of controlling aircraft movement or various aircraft sub-systems. Information automation is dedicated to the task of informing aircrew of the state and position of the aircraft. Management automation assists aircrew in the management of a mission by permitting the pilot to "exercise strategic, rather than simply tactical control over the performance of a mission" (Billings, 1991). Artificial intelligence is capable of supporting/performing many of the pilot functions within automated systems.

Therefore, human-machine interaction is more complex for systems that contain an intelligent component, since such systems are capable of making decisions, generating goals and acting upon them. Such 'executive automation' increases the scope of system automation (Figure 1). A system with executive automation could generate its own goals and pass them to a management automation function which could then monitor the relevant control automation. Note that information automation is not represented in Figure 1 since it does not control a process. Rather, it informs the operator about the process.

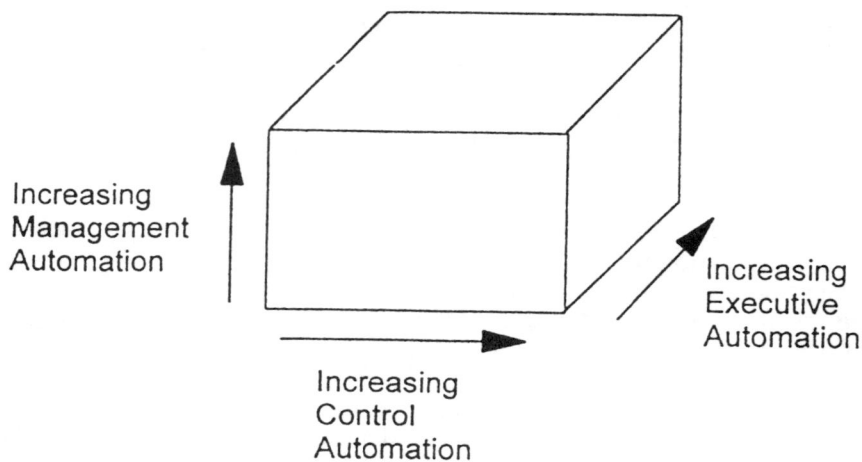

Figure 1. Representation of three 'dimensions' of automation

Sheridan and Verplank (1978) described a set of ten possibilities for human-computer interaction ranging from complete manual control to task execution by a computer system that decides whether the operator needs to be informed of its actions. The latter option raises issues similar to those of adaptive aiding systems, which may transform a task to make it easier, perform part of a task, or allocate whole tasks to the system (Rouse, 1988).

Clearly, it is important to determine the optimal type of collaboration between the human and the IS, based upon effectiveness and safety criteria. However, little or no research has addressed user performance in systems that exploit different levels of automated intelligent support on the basis of adaptive aiding. It seems axiomatic that, if both functions and decisions are automated flexibly, the operator must be aware of the allocation of responsibility in the system. Further, since no system's knowledge representation is perfect (Davis et al, 1993) the IS may try to solve the wrong problem, producing increased operator workload and in turn leading the IS to take over task completion based upon its (faulty) reasoning. A facility is needed that permits the operator to retrieve tasks from the system at any time or to prevent the system from intervening in the first place. Finally, the behaviour of intelligent and adaptive systems is likely to be unpredictable, and may hinder the detection of errors and perhaps lead to low levels of trust in the system (Billings, 1991).

Real-time operations

Many aerospace applications require the monitoring and control of real-time processes. These processes can be described as hard or soft (Stankovic & Ramamritham, 1988). Hard real-time requires that accurate actions be performed within a critical time-frame; failure to achieve this goal may result in disaster. In soft real-time, however, there is still some utility in performing a task after a given time period has elapsed. Interaction design is thus more complicated for hard real-time systems since operators must make very rapid decisions and control responses. Unfortunately, however, much of the human factors research on AI applications may not be pertinent to real-time systems, since a considerable amount of this research has been performed on systems that do not face real-time constraints. For example, aerospace applications are often characterised by the requirement to perform multiple concurrent activities in an environment where the operator has to use a computer to control a process. This is obviously very different from a situation where a doctor's only current task might be using an expert system to diagnose a patient's illness. In such a situation the patient does not even have to be present and the doctor may cease or resume interaction at will. Finally, there are technological problems where AI is used for real-time operations. Research by the USAF has found that rule-based systems may be too slow for real-time tasks (see Laffey, 1991).

Summary

There is no doubt that the use of AI will increase in the future and it seems likely that due to its very nature AI will not be used to perform trivial tasks. However, operators already experience difficulty in determining the current and future actions of automated systems. Extensive use of AI allows the production of systems that are capable of higher levels of autonomy than has previously been the case. The capacity of such systems to define goals and pursue them to their conclusion in an independent manner makes it vital that interaction between the human and the system be designed for maximum safety and efficiency. However, much more research will be required to achieve this goal. This is especially true for adaptive systems, since they may have varying roles, and face considerable communication demands and a number of automation problems.

References

Billings, C.E. (1991), *Human-Centred Arcraft Automation: A Concept and Guidelines*, NASA-TM-103885, Washington, DC.
Card, S.K. (1989), 'Human factors and artificial intelligence', in Hancock, P.A. and Chignell, M.H. (eds), *Intelligent Interfaces: Theory , Research*

and Design, Elsevier, North-Holland, Amsterdam.

Davis, R., Shrobe, H. and Szolovits, P. (1993), 'What is a knowledge representation?', *AI Magazine*, Spring, 19-33.

Fadden, D. M. (1990), 'Aircraft automation challenges' in *Abstracts of AIAA/NASA/FAA/HFS Symposium, Challenges in Aviation Human Factors: The National Plan*, American Institute of Aeronautics and Astronautics, Washington.

Hollnagel, E. (1988), 'Information and reasoning in intelligent decision support systems', in Hollnagel, E., Mancini, G. and Woods, D.D. (eds), *Cognitive Engineering in Complex Dynamic Worlds*, Academic Press, London.

Hoogovens Report (1976), 'Human Factors Evaluation: Hoogovens No. 2 Hot Strip Mill,' Tech. Rep. FR251, British Steel Corporation/Hoogovens, London.

Kessel, C.J. and Wickens, C.D. (1982), 'The transfer of failure-detection skills between monitoring and controlling dynamic systems', *Human Factors*, **24**, (1), 49-60.

Keyes, J. (1989), 'Why expert systems fail', *Expert Systems*, **10**, 3, 151-156.

Laffey, T.J. (1991), 'The real-time expert', *Byte*, January, 259-264.

Madni, A.M. (1988), 'The role of human factors in expert systems design and acceptance', *Human Factors*, **30**(4), 395-414.

Malin, J.T., Schreckenghost, D.L., Woods, D.D., Johannessen, L., Potter, S.S., Holloway, M. and Forbus, K.D. (1991), *Making Intelligent Systems Team Players: Case Studies and Design Issues*, NASA-TM-104738, Washington, DC.

Rouse, W.B. (1988), 'Adaptive aiding for human/computer control', *Human Factors*, **30**(4), 431-443.

Sheridan, T.B. and Verplank, W.L. (1978), *Human and Computer Control of Undersea Teleoperators* (Technical Report) M.I.T. Machine Systems Laboratory, Boston, MA.

Sloane, S.B. (1991), 'The use of artificial intelligence by the United States Navy: Case study of a failure', *AI Magazine*, Spring, 80-92

Stankovic, J. A. and Ramamritham, K. (1988), *Hard Real-Time Systems: A Tutorial*, Computer Society Press, Washington DC.

Wick, M.R., and Thompson, W.B. (1989), 'Reconstructive explanation as complex problem solving', *Proceedings of the Eleventh International Joint conference on Artificial Intelligence*, AAAI, Detroit.

Wickens, C. D. (1992), *Engineering Psychology and Human Performance*, Harper Collins, New York.

Woods, D.D. (1986), 'Cognitive technologies: The design of joint human-machine systems', *The AI Magazine*, **6**(4), 86-92.

Woods, D.D., and Roth, E (1988), 'Aiding human performance: II. From cognitive analysis to support systems', *Le Travail Humaine*, **51**(2), 139-171.

32 Colour and shape coding of head-up display pitch bars

Thomas D. Hardiman, MMI Dept, Defence Research Agency, Farnborough, UK

Abstract

The use of asymmetry in head-up display (HUD) attitude symbology may aid the performance of flight tasks such as recovery from unusual positions (UPs). In a UP, the pilot needs rapidly to ascertain his attitude and recover to the nearest horizon. In the present study, it was hypothesised that the time required for attitude assessment would be shortened by increasing the 'cognitive compatibility' of the HUD. In the conventional HUD, positive and negative pitch are differentiated by the use of solid bars for the former and segmented bars for the latter. Shape and colour were investigated here as additional coding methods. Symmetrical colour and shape were represented by green, tapered bars for both positive and negative pitch. Asymmetry was produced by using blue and/or 'bendi' bars for positive pitch only. Subjects completed four experimental conditions, representing each combination of symmetry/asymmetry of shape and colour. They were required to recover from a series of UPs presented using a flight model on a Silicon Graphics simulator. The performance measures were initial reaction time (IRT) and total recovery time (TRT); subjective workload was also assessed. A significant advantage of colour coding and shape coding was found for IRT, but only when they were presented in combination. Consequently, it was tentatively concluded that HUD attitude symbology should include both colour and shape asymmetry.

Introduction

Traditionally, HUDs have differentiated positive and negative pitch by the use of solid bars for the former and segmented bars for the latter. The present experiment was conducted to investigate the effects of adding shape and colour coding. It was hypothesised that these additional coding methods would facilitate recovery from UPs.

Designing HUDs to aid pilot decision making

Reising, Emerson and Aretz (1984) noted that the introduction of modern technology was changing the pilot's role to one with greater emphasis on decision making. In complex environments, decision making should be rule-based (Rasmussen, 1988). The rule for recovery from a UP is that the pilot should roll and pull to the nearest horizon. Such decision making may be aided by redesigning HUD symbology to provide 'redundancy gain' (improvement in performance when information from one source is integrated with redundant information from an additional source). Selcon, Taylor and Shadrake (1991) demonstrated such an effect when redundant colour coding was used for warning icons.

Designing HUDs for decision making in unusual positions

According to Zenyuh, Reising and McClain (1987), an unusual position may occur from a single factor or a combination of factors. Such factors include turbulence, distraction by cockpit duties, instrument failure and spatial disorientation. Although HUDs are gradually becoming the primary flight instrument in fast-jet cockpits, their inadequacy to display information necessary for recovery from UPs is well documented. Zenyuh et al. (1987) point out that HUDs fail to convey to the pilot that an unusual position exists. Moreover, since the HUD symbology does not clearly indicate whether the aircraft is upright/inverted or climbing/diving, it fails to aid the pilot's decision making concerning recovery using the roll and pull rule.

Reising, Zenyuh and Barthelemey (1988) observed that current HUD attitude symbology does not provide the optimal means of answering the three key questions asked by pilots attempting to recover from an unusual position: 1) am I going up or down?; 2) am I inverted?; and 3) where is the horizon? Taylor (1984) identified the key factors in overcoming such deficiencies as the use of global characteristics that create an easily recognisable overall picture, and exploitation of redundancy gain. In this experiment, redundancy was introduced using shape and colour coding. Zenyuh et al (1987), in a similar experiment, showed that the use of asymmetry through coloured bendi pitch bars contributed to better recovery from extreme UPs. However, in non-extreme attitudes, where the horizon bar was always visible, the standard symmetrical attitude symbology was adequate. Other research (see Dudfield, this volume) on the use of colour coded pitch bars has shown a subjective preference by pilots but failed to find significant performance gains. This may be due to the use of colour coding without shape coding.

Method

Subjects

Sixteen subjects (ten male and six female) completed the experiment. All were volunteers who had no previous experience of flying.

UPs were presented on a HUD display with a white background, using a Silicon Graphics 310 VGX(T).

Experimental design

Each subject performed the four conditions (symmetrical and asymmetrical shape and colour), as depicted in Figure 1. Each condition comprised 72 trials, representing three presentations of each combination of four roll angles (30°, 60°, 120° and 150°) and six pitch angles (70°, 50°, 30°, -30°, -50° and -70°). The order of the conditions was counter-balanced across subjects using a Latin Square, and the order of trials was randomised.

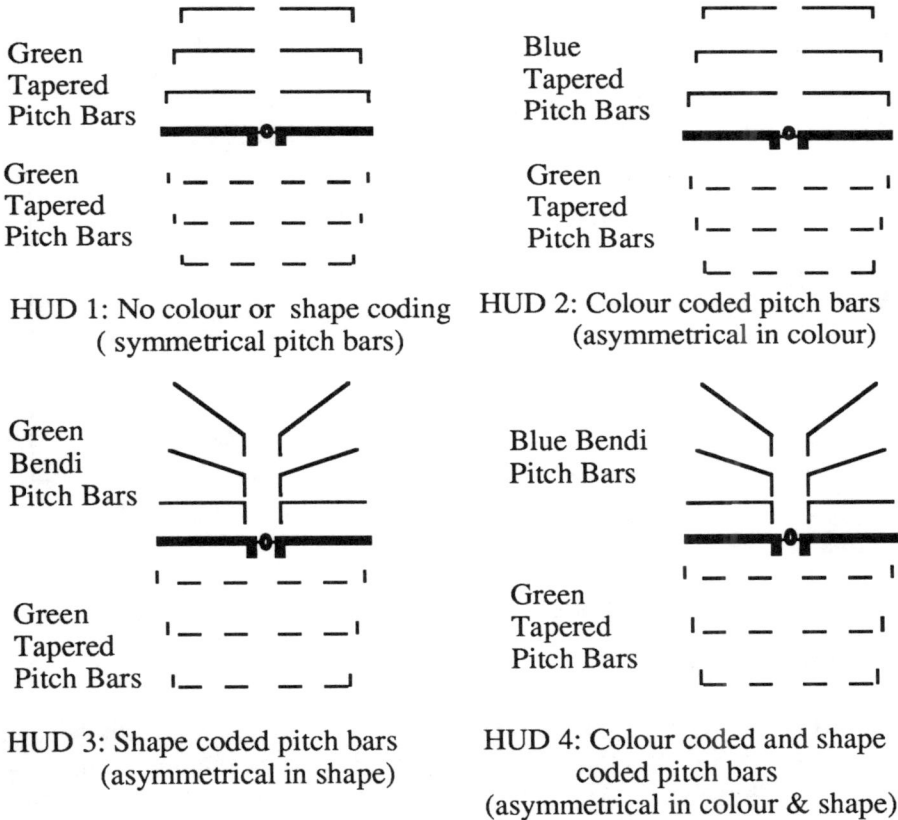

Green
Tapered
Pitch Bars

Green
Tapered
Pitch Bars

HUD 1: No colour or shape coding
(symmetrical pitch bars)

Blue
Tapered
Pitch Bars

Green
Tapered
Pitch Bars

HUD 2: Colour coded pitch bars
(asymmetrical in colour)

Green
Bendi
Pitch Bars

Green
Tapered
Pitch Bars

HUD 3: Shape coded pitch bars
(asymmetrical in shape)

Blue Bendi
Pitch Bars

Green
Tapered
Pitch Bars

HUD 4: Colour coded and shape
coded pitch bars
(asymmetrical in colour & shape)

Figure 1 Symmetrical and asymmetrical HUD symbology

The objective measures were initial reaction time (IRT) in ms (the time taken to make the initial stick movement) and total recovery time (TRT) in ms (the time taken to recover to wings level). Each subject also completed the NASA Task Load Index (TLX) workload questionnaire.

Procedure

In the training stage, subjects were instructed on the use of HUDs and on the correct procedure for recovery from UPs. They were allowed to practise the task until they felt comfortable with the recovery procedure and the format of the alternative HUDs. In the experimental stage, subjects completed all four conditions, providing TLX scores after each condition.

Results

Five-way analyses of variance (ANOVA) were used, with factors of colour coding, shape coding, roll angle, pitch angle and subjects.

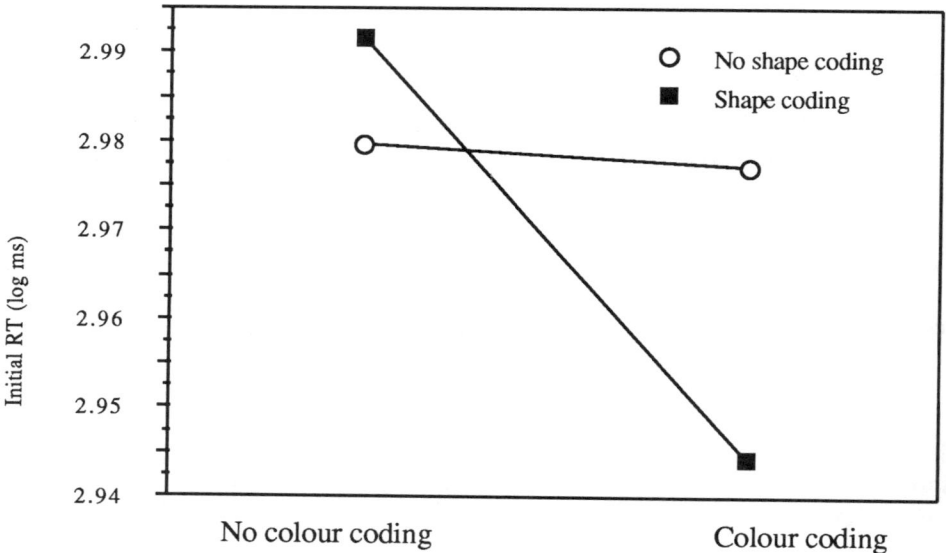

Figure 2 The interaction between colour and shape asymmetry

Initial Reaction Time (IRT)

Subjects' IRT was faster ($F_{1,23}$ = 5.667; p = 0.026) for asymmetrically shaped pitch bars (mean = 968 ms) than for symmetrically shaped pitch bars (mean = 999 ms), and was faster ($F_{1,23}$ = 44.163; p < 0.0001) for colour-coded pitch bars (mean = 953 ms) than for monochrome pitch bars (mean=1013 ms). There was a significant interaction between colour and shape coding ($F_{1,23}$ = 9.276; p = 0.0057), as shown in Figure 2. Post-hoc analysis using t-tests showed that asymmetry of both colour and shape produced faster IRT (mean = 909 ms) than asymmetry only of shape (mean= 1027 ms; p < 0.05), asymmetry only of colour (mean = 998 ms; p < 0.05) and symmetry of both colour and shape (mean = 1001 ms; p < 0.05). Colour and shape coding alone produced no significant advantage relative to the uncoded HUD.

Total Reaction Time (TRT)

TRT did not vary between the four conditions.

TLX scores

Workload scores on the TLX did not differ between conditions.

Discussion

Colour and shape coding of pitch, when present in combination, produced an advantage in IRT. This result supports the concept of redundancy gain, since positive and negative pitch were in all conditions already distinguished on the basis of solid versus segmented bars. TRT, unlike IRT, did not vary between conditions, a finding that may be due to the inexperience of the subjects. The initial advantage may have been lost because of later difficulties in recovering to the horizon. The failure to produce such an overall advantage may underlie the insensitivity of the TLX to differences in the symbology.

Conclusion

These findings provide provisional support for the notion that asymmetry in HUD attitude symbology through the combination of colour and shape coding of pitch bars is desirable. Future research, using pilots as subjects, will be conducted to investigate further the possible operational advantage of these coding mechanisms.

References

Dudfield, H.J. (1994), 'A review of the benefits of colour coding collimated cockpit displays', presented at the 21st Conference of the Western European Association of Aviation Psychologists, Dublin.

Selcon, S.J., Taylor, R.M. and Shadrake, R.A. (1991), *Information Processing Advantages of Multiple Sources of Cockpit Information,* RAF Institute of Aviation Medicine, Report No. 672, Farnborough, U.K.

Rasmussen, J.T. (1987), 'Generic error modelling system: A cognitive framework for locating common human error forms', in Rasmussen, J., Duncan, K. and Leplat, J. (eds), *New Technology and Human Error*, Wiley, Chichester.

Reising, J.M., Emerson, T.J. and Aretz, A.J. (1984), 'Computer generated formats for advanced fighter cockpits', in *Proceedings of the NATO Workshop on Colour-Coded vs. Monochrome Electronic Displays,* RAE, Farnborough, U.K.

Reising, J.M., Zenyuh, J. and Barthelemey, K. (1988), 'Head-up display symbology for unusual attitude recovery', in *The Proceedings of the National Aerospace and Electronics Conference,* Dayton, USA.

Taylor, R.M. (1984), 'Some effects of display format variables on the perception of aircraft spatial orientation', in *Proceedings of the Advisory*

Group for Aerospace Research & Development, Aviation Medical Panel Symposium on Human Factors Considerations in High Performance Aircraft, AGARD, Neuilly-sur-Seine, France.

Zenyuh, J.P., Reising, J.M., and McClain J.E. (1987), 'Advanced head-up display symbology: Aiding unusual attitude recovery', in *Proceedings of the Human Factors Society 31st Annual Meeting,* Human Factors Society, Santa Monica, California.

Part 8
AIRCRAFT MAINTENANCE

33 Maintenance CRM training: assertiveness attitudes effect on maintenance performance in a matched sample

Michelle M. Robertson and James C. Taylor, Institute of Safety and Systems Management, University of Southern California, Los Angeles, California
John W. Stelly, Jr and Robert H. Wagner, Jr, Continental Airlines Inc, Houston, Texas

An airline has adapted its Flight Operations' Crew Resource Management (CRM) program for its Technical Operations organization. This program, called Crew Coordination Concepts (CCC), involves training over 6,500 management and non-management personnel from Maintenance, Planning, Materials, Engineering, and Quality Departments. The CCC training for Technical Operations was offered to management personnel for the first time in June 1991. It is delivered in a two day seminar, facilitated by two instructors drawn from Technical Operations. There are about 20 participants attending each of the training seminars, which are conducted weekly. A high level of interaction among the facilitators and participants is established throughout the training program by using case studies, exercises and role playing. There are six course objectives of the CCC training: 1)identify organizational "norms" and their effect on safety, 2) promote assertive behavior, 3) understand individual behavioral styles, 4) understand and manage stress, 5) enhance rational problem solving and decision making skills, and 6) improve interpersonal skills.

Methods

The maintenance CCC questionnaire development The primary measurement used in this study consists of a questionnaire based on an expansion and modification of a maintenance version of the University of Texas Cockpit Management Attitudes Questionnaire ("CMAQ"). The core questionnaire for the present study contains 28 multiple response items and several open-ended items. A factor analysis of the relationships among the items drawn from the CMAQ confirmed the four attitude clusters reported earlier by Gregorich, et al. (1990). The resulting attitude scales have been labeled "sharing command responsibility," "value of communication and coordination," "recognizing stressor effects," and "avoiding conflict." For this study, the cluster "avoiding conflict" is treated as a reflected scale and labeled "willingness to voice disagreement" to provide an attitude measure of assertiveness. The core questionnaire is used, in five surveys, to measure attitudes before the CCC training, *versus* immediately after the training, as well as two-, six- and twelve- months following the CCC training seminars. The four attitude scales for each of the five surveys are the basic measures of attitudes reported below. Questionnaires that were administered after the training included several open-ended questions which explored how the Technical Operations personnel expected to use what they learned from the CCC training, and how they did use it.

Measures of maintenance performance The trainers and administrators of the CCC course evaluated available performance measures and predicted which of them would be most sensitive to effects of the training. This resulted in five measures from the safety and dependability categories likely to be improved by the training. These performance measures included two safety items (aircraft ground damage and reported occupational injury), and three dependability indicators (departures within 5-minutes, departures within 15 minutes, and delays in planned maintenance due to maintenance error). At the time of writing, a total of 29 months of performance data (6 months before the onset of training and 23 months following) were available for Technical Operations.

Analysis processes For the present analysis we have chosen two (of the five) performance measures and a sample from one of the four after-training surveys, to illustrate the relationships between assertiveness (a specific, intended CCC training effect) and maintenance performance.

Results

We will describe some of the overall findings related to changes in maintenance performance for selected safety and dependability measures, as well as perceived attitude and behavioral changes following the CCC training.

Maintenance performance before and after the onset of training

Safety The first performance indicator, occupational injuries, is shown in Figure 1, which shows these safety results for six months before and the two years after training. The pattern of results shows the directional trend for each of the two periods separately. It is evident that the pattern after training is lower (better safety) overall, with the exception of two months during Summer 1991.

Dependability An important measure of dependability is departures within five minutes. Figure 2 displays the trends for this index for six months before, and the two years after, the training began. Dependability shows steady continued improvement in pre- and post-training performance. Although the training seems to have no <u>overall</u> effect on this measure it is shown below to be related to some attitude changes -- particularly assertiveness.

Training effects on attitudes and perceptions

Comparison of attitudes pre- and post-training and at two month, six and twelve month follow-up Figure 3 presents the comparisons for the four attitude scales from the total population of usable questionnaires returned. Attitude changes from pre-training to post-training were analyzed, and for three of the four attitude scales a significant increase was found immediately following training. No significant difference was found for the "willingness to voice disagreement" scale immediately following training. That attitude measure of assertiveness was significantly higher between the post measure and the 2-month follow-up survey. All four attitude scales remained high, and stable, at two, six and twelve months following the CRM training.

How training was used on the job For the participants who returned the two, six and twelve month follow-up questionnaires, responses to the question of how they actually used the CCC training on the job were content coded. The trainees' self-perception of behavior responses fell into three categories: 1) "Better listening," 2) "More awareness of others," and 3) "Dealing better with others." These categories represent two types of self-perception of potential behavior change. The first two categories show a "passive" improvement made within the person, while the last category shows an "active" response by direct interpersonal approaches. Figure 4 shows how these three categories of training use were reported two, six and twelve months after training. The percentage of respondents reporting "better listening" tended to decrease over time, while the other passive category, "be more aware of others," showed a more stable pattern over the three survey periods. The active behaviors included in "deal better with others" rose substantially at twelve months. A statistically significant difference was found among the patterns over time. The preferred behaviors shifted from those people could do by themselves (e.g., "be a better listener" and "being more aware of others"), to behaviors which actively involve others, (i.e., "dealing better with others").

217

Examining the shift toward assertiveness and its relationship with performance
Figure 3 showed that "willingness to voice disagreement" did not increase
appreciably until *two months after* the CCC training. It was also noted above
that preferences for more active interpersonal behaviors (a logical operationaliza-
tion of "assertive" actions) also increased in the survey two months following
the training.

To more closely examine the possible effects of assertiveness on maintenance
performance we have drawn the sample of 149 2-month follow-up questionnaires
representing those managers whose surveys could be matched to their earlier
surveys by ID code and who had completed training between June and December
1991. These questionnaires were returned by mail between October 1991 and
February 1992. Selecting that sample of respondents permits the examination
of correlations between 2-month follow-up attitudes and maintenance unit performance for a *substantial* number of months (15 months) following that survey.
Thus the sample selected offers a clear picture of the relationship between
managers' improved attitudes and the subsequent performance of their
subordinates. The use of such time-lagged performance allows added certainty
to causal inferences about attitudes and behaviors on performance.

Relationship between attitudes and safety performance (Occupational Injury)
Of the four attitude scales only "willingness to voice disagreement" was found
to be positively related to lower reported occupational injury in numbers of
correlations significantly greater than we would expect by chance alone. That
attitude scale is significantly related to safety for three of the six months before
the training; to two of the three months following training but prior to the follow-
up survey; and for six of the 15 months following the survey for this sample.
These findings suggest that the more their subordinates exhibit favorable safety
performance, the more managers (who may already show some inclination to
act assertively),in all Technical Operations departments, will recognize the value
of speaking up and value encouraging others to do so too. The opportunity
to practice assertiveness in the months following their training apparently is
validated for the managers, as they see evidence of their newly developed skills.

*Relationship between attitudes and dependability performance (Departures within
5 minutes)* Like safety, this dependability measure is strongly and positively
related to "willingness to voice disagreement." That attitude scale is related
to dependability for each of the six months before the training; to two of the
three months following training but prior to the follow-up survey; for all five
months during that survey, and for 10 of the 15 months following the survey
for this sample. This indicates that initial success in dependability encouraged
line station managers with a tendency toward assertiveness to see that quality

218

as useful and led to further post-training commitment and behavioral change in the direction of speaking up and encouraging others to be assertive, which in turn led to even better on-time performance. A pattern of significant and positive correlations between dependability and "recognition of stressor effects" is found for three of the months during the survey, and for nine of the 15 months after the survey. The interpretation for this pattern suggests that the training made managers aware of the effects stress has on their performance. For those managers with the highest and most lasting impressions of stress effects after the training, they translated that feeling into better stress management, which led to improved dependability performance at their stations.

Conclusion

Specific measures of safety and dependability performance show somewhat different responses toward improvement following CRM training. However, these two performance measures show similar patterns of relationships with positive attitude toward assertiveness before and after training. This is persuasive evidence that the training affects performance, even though the training effects operate through indirect linkage. Performance success together with the training influences the managers attitudes toward assertiveness and the managers', in turn, influence their mechanics who, in turn, perform more safely and with greater dependability. Many of the managers in the sample examined report changing their behavior in the months following the training to take full advantage of what they have learned. Additional survey data continue to be collected in the months following the training, thus larger samples sizes and longer time periods will become available for further analyses. The company has begun to provide recurrent CCC training for maintenance managers and to adapt the CCC program for training mechanics as well. Based on the results presented here, we expect to see even stronger evidence for the power and effectiveness of the CCC training in improving safety in the future.

Notes

1. This research is funded under research grant NCC 2-812 from NASA Ames Research Center, and credits both NASA Office of Space Science and Applications and the Aerospace Human Factors Research Division of Ames Research Center. It is also supported by Continental Air Lines.

References

Gregorich, S.E.; Helmreich, R.L. & Wilhelm, J.A. (1990). 'The structure of cockpit management attitudes', *Journal of Applied Psychology*, **75** (6), 682-690.

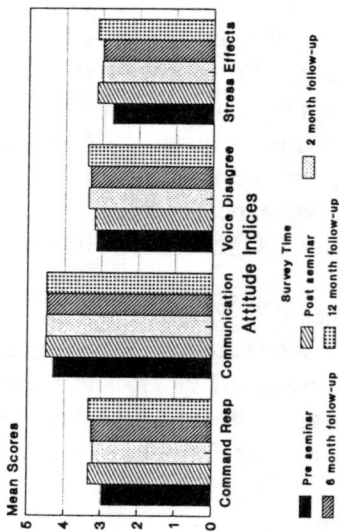

FIGURE 1 SAFETY
OCCUPATIONAL INJURY
Time Series

—○— PRE TRAINING   —×— POST TRAINING

Incidence rate per 1000; 3/94

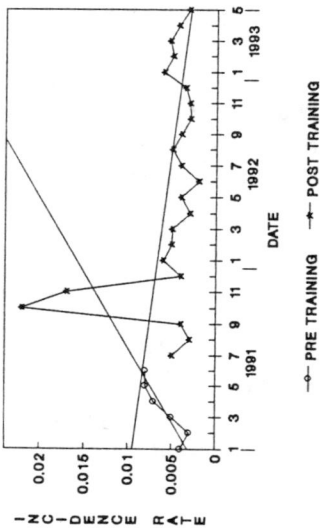

FIGURE 2 DEPENDABILITY
DEPARTURE IN 5 MINUTES
Time Series

—○— PRE TRAINING   —×— POST TRAINING

3/94

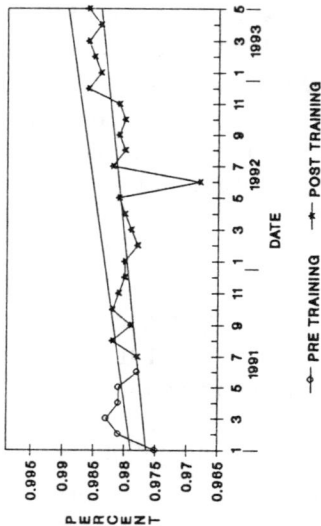

FIGURE 3 ATTITUDE INDICES
Pre, Post, Two, Six, Twelve Months

Survey Time

■ Pre seminar   ▨ Post seminar   ▦ 2 month follow-up
▧ 6 month follow-up   ▤ 12 month follow-up

pre/post n=1996; 2n=734; 6n=636; 12n=413

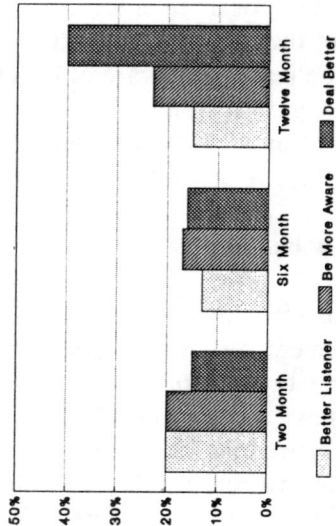

FIGURE 4 REPORTED USE OF THE TRAINING
How Training Was Used on the Job for
Two, Six, Twelve Month Follow-up

□ Better Listener   ▨ Be More Aware   ■ Deal Better

2 mos, n=285; 6 mos, n=192; 12 mos, n=84

220

Part 9
PHYSIOLOGICAL FACTORS

34 Pilots, performance and hypoxia: a review of some research into the effects of mild hypoxic hypoxia

Robert Henderson[1] and Dianne McCarthy, Department of Psychology, The University of Auckland, New Zealand
Ross St George, Department of Psychology, Massey University New Zealand

Introduction

The study of the effects of hypoxia on human physiology and behaviour has accompanied aviation since its inception with ballooning and predates both fixed and rotary wing flight. In 1875, Tissandier, Croche and Sivel ascended to 8784m in a balloon equipped with crude oxygen equipment. Although they had been trained to use the equipment, they delayed using the oxygen to conserve their supply. Hypoxia overcame them, they lapsed into unconsciousness, and when the balloon eventually returned to the ground only Tissandier was alive.

By contrast, in 1993 Harry Taylor completed the 8848m ascent of Mount Everest without oxygen. He achieved this through careful acclimatisation over an extended period. The effects of hypoxia are different for flight crews and their passengers. Because of the rate at which their ascent is made, there is often insufficient time available for the acclimatisation necessary for survival and well-being at altitude.

Definition of hypoxia

Living organisms obtain energy by oxidising complex chemical foodstuffs into simpler compounds. Oxygen is a critical element in this process. The absence of an adequate supply of oxygen to the tissues, whether in quantity or molecular concentration, is termed *hypoxia*. Humans are vulnerable to the effects of oxygen deprivation and severe, or acute, hypoxia nearly always results in a rapid deterioration of body functions. *Hypoxic hypoxia,* one of the four forms of tissue hypoxia, arises from unprotected exposure to altitude, and is caused by the absence of an adequate supply of oxygen in the arterial blood and hence capillary blood. The physiological consequences of hypoxic hypoxia affect all those who aviate, crew and passengers alike.

Effects

A comparison of percentage arterial oxyhaemoglobin saturation (SaO_2) versus altitude, shown in Figure 1, illustrates the rapid decrease in SaO_2 that occurs above 3048m. The physiological effects of hypoxia in aviation are well understood in terms of gaseous exchange, times of useful consciousness and physical reactions, and excellent summaries are available in the literature (e.g., Ernsting and King, 1988).

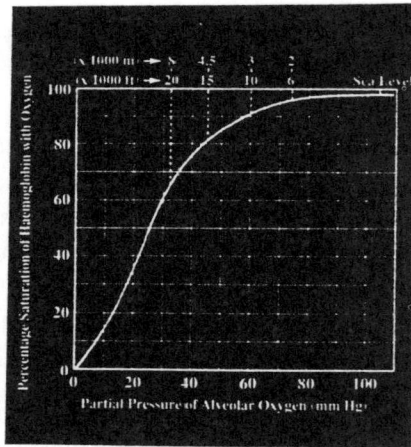

Figure 1   Carriage of oxygen in the blood by haemoglobin
This shows that oxygen tension falls rapidly above 3000m (10,000 ft)

Exposure to hypoxic hypoxia produces visual disturbances, general physiological and cognitive symptoms, and specific neuro-muscular effects as detailed in Table 1. Physiologically a person exposed to hypoxic hypoxia below 3048m experiences a reduction in their SaO_2 level from a sea level value of about 98% to a value of approximately 92 to 93%. In response to this relatively mild insult, the respiration and cardiac rates increase slightly. There are, however, two replicable performance deficits from exposure to mild hypoxia: the dark-adapted eye suffers reduced sensitivity as low as 1219m, and performance on novel tasks is usually impaired at altitudes of about 2438m and higher. Physical activity, a cold environment, illness, and the ingestion of certain drugs, including alcohol and tobacco, all modify an individual's response to hypoxia. As altitude increases the various signs and symptoms are accentuated and the time required for their onset reduces.

Table 1
Effects of exposure to hypoxic hypoxia

Visual	General	Neuro-muscular
Decrease in colour perception	Euphoria	Clumsiness
Decrease in peripheral	Task fixation	Fine tremor
awareness	Personality changes	Slurring of speech
Decrease in acuity	Fuzziness (not dizziness)	Slow movements
Dimming	Amnesia	Hypoxic 'flap'
	Lethargy	
	Mental confusion	
	Sensitivity to cold or heat	
	Cyanosis	
	Loss of self criticism, judgement	

The aviation working environment

The aviation working environment constantly exposes pilots, crews, and passengers of modern aircraft, working at cabin altitudes of up to 3048m, to mild hypoxia for prolonged periods of time. In addition, pilots involved in general aviation, recreational flying activities, and flying training also suffer from the effects of mild hypoxic hypoxia.

While hypoxia has long been recognised as having detrimental effects on human performance, there are conflicting reports in the literature regarding its psychological effects at operational altitudes up to about 3658m. International aviation regulations dictate supplementary oxygen for aircrew operating at altitudes exceeding 3048m. However, various researchers have recommended differing altitudes at which supplementary oxygen should be available, with scientific cases being made for supplementary oxygen to be used anywhere between 1219 and 3658m. Civil regulations, by comparison, may permit flight above 3048m, and up to 4267m for 30 minutes, without oxygen.

Since the 1960s, commercial aviation has witnessed a substantial reduction in the number of accidents per passenger mile, but the percentage of accidents ascribed to pilot error has remained constant. Many instances of pilot error can be traced back to a lack of situational awareness: either the crew misinterpreted the information with which they were presented, or the speed and complexity of events allowed too little time for complex decision making. The United States National Transportation Safety Board has concluded that flight-deck crew performance was a key factor in 37 major accidents between 1978 and 1990. Procedural errors, poor decision-making and monitoring failures were identified as specific contributors and Learmount (1993) has recently criticised modern flight decks, relying on the electronic presentation of information, as being ergonomically complex and mentally 'indigestible'.

Contradictions

A number of contradictory results are present in the literature. For example, Critical Flicker Fusion Frequency (CFFF) thresholds have been found to both decrease and to be unchanged on exposure to hypoxia (Tune, 1964), while the thresholds have also been found to be sensitive to age, light-dark ratios, and the background luminance (McFarland, 1958). Fiorica, Burr and Moses (1971) reported their sea level group, breathing 100% oxygen during the latter part of a four hour vigilance task, missed more signals than their hypoxic experimental group at 3505m. Vaernes, Owe and Myking (1984) found various performance decrements during an extended 6.5 hour experiment at 3048m but no relationship between impaired performance and duration of exposure. For example, reasoning was impaired initially and also after two hours, but performance recovered after four hours. However, others (e.g., Crow and Kelman, 1973) report no decrement below 3658m on tasks involving short-term memory and Green and Morgan (1985) report no effect of altitude on a logical reasoning task at 3660m.

Various results have been reported for Reaction Time (RT) thresholds under conditions of mild hypoxia. These thresholds range from 1524m (Denison, Ledwith and Poulton, 1966), through 2438m (Farmer, Lupa, Dunlop and McGowan, 1993), 2972m (Fowler, Elcombe, Kelso and Porlier, 1987) and 3048m (Tune, 1964), while Ledwith (1970) has reported faster RT's at both 4267m and sea level relative to 2134m.

Confounding variables

The experiments cited serve to illustrate the variability in performance measures when subjects are exposed to mild hypoxia. The failure to find consistent effects of hypoxia at operational altitudes may be due to four important factors: (1) the use of naive subjects with no previous exposure to hypoxia where their ignorance creates apprehension (Ledwith, 1970; Green and Morgan, 1985); (2) individual variations in the physiological response to hypoxia with the critical independent variable being the 'effective' altitude of the subject (as measured by their SaO_2 level) rather than their 'nominal' altitude (as measured by the altimeter) (Fowler et al, 1987; Farmer et al, 1993); (3) variations in the duration of exposure to altitude with previous experiments, with few exceptions, lasting between 30 minutes and 1 hour; and, (4) the use of tasks with little relevance to those performed in the aviation environment. For example real instrument tasks, such as ILS approaches, have not been used to monitor performance in hypoxia research.

The way ahead

The overall aim of the project being undertaking is to measure the performance of qualified pilots under conditions of reduced blood-oxygen saturation levels. The experiments will be conducted in the Royal New Zealand Air Force (RNZAF) hypobaric chamber and the performance of pilots on a battery of tests will be measured while they are exposed to controlled levels of hypoxia between sea level and 4267m. Specific performance measures will include PC-based simulated flying tasks, and a variety of reaction time, memory, and attentional tasks. Additionally, 'effective' altitude will be measured from arterial blood samples obtained at 3048m. The use of qualified pilots as subjects will minimise apprehension; the individual SaO_2 responses will be monitored throughout the experiments; the exposure duration will be six hours, and performance will be assessed on tasks which closely approximate those required of a pilot.

Conclusion

Despite the contradictions in the literature, there do seem to be psychological effects present in the altitude range at which commercial and recreational pilots operate in either pressurised cabins or unpressurized aircraft. Perhaps more importantly, training flights operate in the 2438 to 3048m band and the one result that does keep appearing in the literature is that mild hypoxia affects the learning of new tasks. This result has both training and economic implications. While some psychological research has reported impaired cognitive performance (e.g., reaction time, memory, attention) at altitudes between 1524 and 2438m, other experimenters have failed to show consistent and reliable performance deficits below 3658m. The role of hypoxia, therefore, as a causal factor in aircraft accidents is less than clear and the empirical identification of hypoxia induced performance and memory decrements will contribute to the enhancement of aircraft safety for aircrews and passengers.

Notes

1. Former Wing Commander, RNZAF; now Doctoral Student; University of Auckland.

References

Crow, T.J. and Kelman, G.R. (1973). Psychological effects of mild acute hypoxia. *British Journal of Anaesthesia.* **45**, 335-7.

Denison, D.M., Ledwith, F. and Poulton, E.C. (1966). Complex Reaction Times at Simulated Cabin Altitudes of 5000 Feet and 8000 Feet. *Aerospace Medicine,* **37**(10), 1010-13.

Ernsting, J. and King, P. (Eds) (1988). *Aviation Medicine 2nd Ed.* London, Butterworths.

Farmer, E.W., Lupa, H.T., Dunlop, F. and McGowan, F.F. (1993). *Task learning under mild hypoxia.* Unpublished IAM Farnborough Report.

Fiorica, V., Burr, M.J. and Moses, R. (1971). Effects of Low-Grade Hypoxia on Performance in a Vigilance Situation. *FAA Office of Aviation Medicine Report No. AM-71-11.*

Fowler, B., Elcome, D.D., Kelso, B. and Porlier G. (1987). The Threshold for Hypoxia Effects on Perceptual-Motor Performance. *Human Factors,* **291**, 61-6.

Green, R.G. and Morgan, D.R. (1985). The Effects of Mild Hypoxia on a Logical Reasoning Task. *Aviation, Space, and Environmental Medicine,* **56**, 1004-8.

Learmount, D. (1993). The limiting factor. *Flight International,* **1444**(4393), 40-1.

Ledwith, F. (1970). The effects of hypoxia on choice reaction time and movement time. *Ergonomics,* **13**, 465-82.

McFarland, R.A., Warren, A.B. and Karis, C. (1958). Alterations in critical flicker frequency as a function of age and light:dark ratio. *Journal of Experimental Psychology,* **566**, 529-38.

Tune, G.S. (1964). Psychological effects of hypoxia: Review of certain literature from the period 1950 to 1963. *Perceptual and Motor Skills.* **19**, 551-562.

Vaernes, R.J., Owe, J.O. and Myking, O. (1984). Central Nervous Reactions to a 6.5-Hour Altitude Exposure at 3048 Meters. *Aviation, Space, and Environmental Medicine,* Oct, 921-6.

35 Illness or incapacitation in aviation safety incidents

Rudolf G. Mortimer, University of Illinois, Champaign, Illinois

In a study of loss of consciousness due to cardiovascular and neurologic conditions among U.S. air force pilots (McCormick and Lyons, 1991) the rate of incapacitation was 0.19 per million flying hours. This rate predicts 5 and 6 incapacitating events per year, respectively, in general and commercial aviation, and is similar to other studies (Mohler & Booze, 1978; Mortimer, 1991).

Other forms of illnesses that occur during flight are less well documented whether occurring among air crew or passengers.

In a review of preexisting diseases found in 809 military, professional and private pilots killed in flying accidents in the United Kingdom between 1955 and 1979, the most common ones were coronary atherosclerosis, myocarditis, liver pathology, upper-respiratory tract infections, defective vision, and psychiatric or adverse medical histories (Underwood Ground, 1981).

A survey of in-flight medical emergencies in business aircraft was made by Garrison (1991). Between 1971 and 1989 there were 3.4 incidents per year and a fatality rate of 0.19 per 100,000 hours flown. About 69% of the incidents involved passengers and the remainder were crewmembers.

Rotenberg (1987) found that medical incidents per passenger were about 4.76 and 7.69, respectively, per 100,000 passengers as reported by Qantas and British Overseas Airways. Another study (Cummins and Schubach, 1989) showed that there was a rate of 2.52 en-route medical incidents per 100,000 inbound passengers at the Seattle-Tacoma International Airport.

In a survey of pilots (Mortimer, 1993) the most common illnesses reported were headaches, earaches, motion sickness, joint and muscle pain, chest pain, stomach ache, visual impairments, colds and sinusitis. About 20% of the pilots reported sleeplessness or fatigue and 60% of those felt that it adversely affected their performance. Also, about 28% reported stress due to home, aircraft condition, flight condition or air traffic control.

Thus, although there have been studies made of the types of illnesses and incapacitations that occur to aircrew and passengers in general aviation and commercial aviation flights, there is relatively little known of the incidents that resulted that had an impact on flight safety.

Method

The data for this study were from NASA's Aviation Safety Reporting System. A total of 356 reports were obtained out of a total database of 38,051 full-form records that were received between January 1, 1986 and December 1992. Each case contained a narrative description from which the basic information that was used was coded. There were 199 useable reports.

Results

Individual affected and reporter of the event

The person making an ASRS report is not necessarily the person causing the incident. Among aircrew, the pilot flying made 28% of all reports, crewmembers not flying made 33%, and cabin crew 1%. General aviation (GA) pilots made 27% of all reports and ATC made 8%. Flight instructors made 3% of all reports about themselves, student pilots and a passenger. The reports about passengers were all made by aircrew or general aviation pilots, except one case reported by ATC. Overall, the pilots flying made 55% of the reports.

Overview of findings

About 43% of the reports involved slight, 34% partial and 19% total incapacitation. Aircrew were 42% of those affected, GA pilots were 17%, passengers 29%, air traffic controllers 8% and cabin crew about 4%.

The most common flight incidents were infractions of FAA or company procedures (32%), cruise, climb or descent altitude deviations (20%), route deviations (7%), airspace and traffic pattern violations (9%), approach course or altitude deviations (6%), aircraft handling (6%) and ATC routing and altitude errors(5%).

About 23% of the incidents had some roots in personnel conflicts, mostly between management and employee. For example, a crewmember reported for duty while ill for fear of management reprisal if calling in sick. About half the incidents occurred at altitudes above 10,000 feet. An IFR flight plan had been filed in 77% of cases and VFR in 6% and none in the rest; 94% of the pilots flying were instrument rated, 86% were ATP or Commercial certified and 1%

230

were students.

Weather, visibility restrictions or runway conditions were contributing factors in only a small number of cases. Visual meteorological conditions prevailed in 78% of cases.

Numerous kinds of illnesses were reported as well as fatigue, stress and anxiety. Most common were headache, flu, cold and "illness" (27%), gastro-intestinal problems (17%), fatigue (10%), cardio-pulmonary, abdominal pains or hemorrhage (11%), vision, hearing, back or muscle impairments (11%), airway obstruction (7%), anxiety (6%) and hypoxia or carbon monoxide poisoning (5%).

Total incapacitation

About 20% of the incidents involved total incapacitation, of which 49% were aircrew, 46% passengers and two cases GA pilots. Both the latter cases were caused by hypoxia. Other causes of total incapacitation were flu, cardiopulmonary, hemorrhage, diabetes, gastrointestinal, abdominal, anxiety/hyperventilation, backache, burns, neurologic and airway obstruction. One third of the events were due to cardiopulmonary and gastrointestinal problems.

Slight v. partial/total incapacitation

Passengers were more often stricken with partial or total incapacitating causes (79%) than aircrew (48%) or GA pilots (46%) or ATC (20%).

Phase of Flight

About 14% of the incidents happened prior to or during take-off. Most incidents began in the cruise phase (46%), but 30% in approach and 10% in landing and after--the later stages of a flight.

Type of incident

There was a significant association between the degree of incapacitation and type of flight incident. In particular, when the incapacitation was partial or total there were significant increases in procedural errors while those with "slight" incapacitation involved more deviations from altitude or speed restrictions and approach altitude or course deviations.

Type of incident by persons affected

When the person affected was a member of the aircrew there were more

231

violations of altitude or speed restrictions and of procedures; if a passenger, more airspace and traffic pattern violations; if a GA pilot, more taxi, route, radio procedures, approach altitude and course deviations and airspace and pattern violations.

Altitude

There was a significant association between degree of incapacitation and flight altitude. Of those totally incapacitated, 73% occurred at 10,000 feet or greater. By comparison, 40% and 41%, respectively, of those slightly or partially incapacitated occurred above 10,000 feet.

Passenger v. other operations

About 64% of the incidents involved air carriers. There was a significant association between the types of flight incidents and the type of operation. Air carriers had relatively more altitude or speed violations and procedural errors. Others had more incidents involving improper aircraft handling, approach course and altitude deviations, and airspace and traffic pattern violations.

Incidents by individual affected

Aircrew: Total incapacitation of aircrew was primarily due to cardiopulmonary, gastrointestinal, and abdominal problems, which accounted for 65% of the causes. Fatigue was a factor in 15% of the incidents which involved slight or partial incapacitation, while colds, flu, laryngitis and like illnesses were 28% of all. The total and partial incapacitations mostly resulted in procedural errors, while slight incapacitations mostly produced altitude or speed deviations.

General aviation pilots: There were only two cases of total incapacitation among GA pilots, both due to hypoxia. The most common causes of slight or partial incapacitation were due to cold, flu, "illness" (18%), fatigue (12%), disorientation/vertigo (15%), carbon monoxide poisoning (9%), foreign substance in eye (9%), smoke/fume inhalation (9%), and anxiety, dehydration, low potassium and gastrointestinal problems.

These incapacitations mostly contributed to airspace and traffic pattern violations (28%), taxi, route or radio procedure errors (22%), approach course/altitude deviations (16%), improper handling (13%), improper procedures (13%) and altitude deviations (9%).

Passengers and cabin attendants: Incapacitation was reported among passengers and cabin crew in commercial (60%) and general (40%) aviation and were 29% of the incidents, of which 31% were described as totally incapacitating. The latter were due to a variety of factors: flu, headache, hemorrhage, cardiopulmonary, diabetes, anxiety, airway obstructed, and a burn to a cabin attendant. Among the slight or partial incapacitating events, the most common were gastrointestinal (55%), cardiopulmonary (10%), cold, flu (13%) and hemorrhage, abdominal, obstructed airway, hypoxia and CO poisoning.

The total incapacitations mostly resulted in violations of assigned altitude or speed (38%) and procedures (38%), whereas the slight or partial incapacitations mostly resulted in airspace and traffic pattern violations (26%), procedure violations (26%) and approach altitude and course deviations (15%).

Air traffic controllers: About 8% of reports were submitted by controllers who reported slight or partial incapacitation due to cold, flu or unspecified illness, headache, fatigue and effects of prescription medicines. These resulted in procedural and routing errors, which mostly led to altitude conflicts and were 56% of all ATC incidents. The controller was the person incapacitated in all but one of the incidents. That case involved an aircraft passenger suspected of a heart attack which led to an altitude conflict due to the controller's erroneous instructions to the aircrew.

Discussion

Two GA pilots suffered total incapacitation. One report was made by the flight instructor who was with the student pilot who lost consciousness. The other was made by the lone pilot who was lucky to regain consciousness from hypoxia as his airplane lost altitude. This leads one to wonder about other cases of total incapacitation in single pilot operations. In this sample of ASRS incidents total incapacitation affected 22% of aircrew and only 6% of GA pilots, while 52% of aircrew and 55% of GA pilots were slightly incapacitated. If the probability of total incapacitation among aircrew and GA pilots is assumed to be proportional to their slight incapacitation, as seems reasonable, then it would be expected that about 21% of GA pilots would have had totally incapacitating incidents--about three times more than were reported.

The kinds of incapacitations that were reported by aircrew and GA pilots were similar to those obtained in a survey of pilots (Mortimer, 1993), which lends credence to both studies. In both studies fatigue, cold, flu, headaches, back and muscle aches and anxiety were commonly mentioned.

There were some differences in the ASRS reports of aircrew and GA pilots. While 15% of GA pilots reported disorientation/vertigo, only 6% of aircrew were affected. Conversely, GA pilots reported cardiovascular and gastrointestinal problems in 6% of incidents, while 20% of aircrew reports mentioned these problems, which were often totally incapacitating. These findings reflect an inherent disadvantage of self-reports, but they suggest, not unexpectedly, that solo GA pilots are at greater risk of accidents due to physical or other incapacitations. Control of cardiovascular problems in aviation is difficult to achieve beyond the medical exam already required and emphasizes the importance of striving for good general fitness by pilots. Many of the other health factors that were reported are recogniz able by pilots and were noticed prior to flight. Other factors affected the decision to fly. Crewmembers sometimes mentioned that they were under pressure to report for duty in spite of their illness. The same was true of air traffic controllers.

Perhaps this study will help to show the need to consider more carefully the potential consequences of illnesses and incapacitaions for flight safety.

Acknowledgements

This research was supported by the U.S.Army Construction Engineering Research Laboratory. Research assistants were Mr. Yang Li and Mr. David Gill. The data were made available by Dr. Sheryl L. Chappell (NASA) and Ms. Stephanie M. Frank (ASRS-Battelle).

References

Cummins, R.O. and Schubach, J.A. (1989),'Frequency and types of medical emergencies among commercial air travelers. *J. Amer. Med. Assoc.*, 1295-1299.

Garrison, R.T. (1991), 'Incidence of in-flight medical emergencies in business aircraft', MS thesis, Dept. Community Health, Wright State University.

McCormick, T.J. and Lyons, T.J. (1991), 'Medical causes of in-flight incapacitation: USAF experience, 1978-1987', *Aviat. Space & Env. Med.*, **62**, 884-887.

Mohler, S.R. and Booze, C.F. (1978), 'U.S. fatal general aviation accidents due to cardiovascular incapacitation: 1974-1975', *Aviat. Space & Env. Med.*, **49**, (10), 1225-1228.

Mortimer, R.G. (1991), 'Some factors associated with pilot age in general aviation crashes', Proc., 6th International Symposium on Aviation Psychology, Columbus, OH. 770-775.

Mortimer, R.G. (1993),& 'Illness, drugs, fatigue and stress in the cockpit reported by pilots', Proc., 7th International Symposium on Aviation Psychology, Columbus, OH.

Rotenberg, H. (1987), 'Medical emergencies aboard commercial aircraft', *Annals of Emergency Medicine*, 1373-1377.

Underwood Ground, K.E. (1981), 'Occurrence of pre-existing disease in air crew killed in flying accidents', *Aviat. Space & Env. Med.*, **52**, 672-676.

36 Relationships between aviation physiology and aviation psychology

Joseph L. Vogel, Adjunct Assistant Professor, The Ohio State University

Introduction

Humans are terrestrial creatures, bound to the surface of the earth, and destined for disaster if they depart too far from that homeland. Humans are also thinking, reasoning creatures, capable of solving problems and rising above their station in life. For centuries, men were told not to rise above their station lest they be brought down by the wrath of God, or by some other unknown disaster. For years people who studied the natural world were ostracized, pilloried, or killed. A cruel fate awaited those who went against the accepted authority.

Yet, despite the threats, the conventional wisdom, and the wrath of the authorities, some people did go against the accepted thinking of the era, some sought answers outside the teachings and writings of the day, and some even rose above their station - far above! It was inevitable that along the way to discovery, some lost their way, some lost their lives, and some found new answers to old problems. In retrospect, some of those answers were quite simple. They came because a person asked the right question, or questioned the answers that were already given, or because they had the temerity to combine two or more ideas or disciplines that had never before been combined.

The Wright brothers questioned the tables of pressure compiled by Lilienthal and found them to be wrong. They were inspired by Octave Chanute, yet they questioned the wisdom of this accepted and even revered scientist. The Wrights accepted Chanute's bridge building talents and used his ideas to build a strong biplane structure. They, in-turn, instinctively saw that Lilienthal's weight shift method of control was flawed and sought controllability over natural stability. Because they dared to go their own way and had the good sense to combine ideas from other sources, they were ultimately successful in conquering the realm of controlled, powered flight.

What I intend to propose may put me in the same category as those medieval thinkers who went against the conventional wisdom and offered a new way to look at old problems. But before that proposal is outlined, it would be well to review what the physiological and psychological effects are that act upon the body and brain of an aircrew member in flight.

Physiological effects

The physiological effects of oxygen deprivation, hypoglycemia, poor diet, lack of physical fitness, stress, fatigue, and other causes are well known to military aviators, test pilots, and some airline pilots. They are less well known to civilian pilots due to lack of training. These and other effects are certainly not very well known at all by persons outside the discipline. The aviation physiologist strives to find ways to protect the aviator from the hazards of flight beyond the so called "Physiological Zone" which has been designated by the (USA) Federal Aviation Administration as the area from sea level up to about 12,000 feet. This is the area in which the human body is more or less adapted. However, unless humans become acclimatized, such as the natives of Peru or the Andes Mountains, lengthy stays, even in this area, can be detrimental to them in a physiological sense. Flying higher than the physiological zone without protective equipment and without supplemental oxygen can lead to disaster.

Psychological effects

Human beings bring to aviation a variety of traits, beliefs, and attitudes that bear heavily upon their performance. Basically these traits, beliefs and attitudes cannot be used to directly determine how any one person will react to any one specific situation. However, one can infer from certain traits that people will generally behave in a set pattern of behavior. The study of human factors strives to find the causes for human behavior in a general sense and attempts to apply the lessons learned in a real-time sense.

When aviation was new and aircraft were just being invented, every manner of control configuration was tried. The Wright brothers even had aircraft with "Orville" controls and aircraft with "Wilbur" controls. When one views these two-stick, side-to-side and fore and aft controls from a current perspective, it is a wonder that they were able to control the aircraft at all! It took France's Louis Bleriot eight tries to come up with the center stick and rudder bar controls that we consider to be "modern." It was not until the Second World War that the Psychology Laboratory at Cambridge University began to study the problem. The conclusion was that displays and controls should be optimized for and match the capabilities of the person instead of making the person conform to an uncomfortable and inefficient cockpit.

Engineers want to place switches and controls where it is convenient in manufacture. Financial planners want them where they will cost the least money and sales people want them where they will enhance the look and feel of an aircraft, leading more people to buy. Human Factors researchers want controls and switches placed where a human operator will naturally reach for them in the context of a normal flight.

Accident causes

Of all of the possibilities that can be blamed for aircraft accidents, including mechanical failure, only one stands out clearly as number one. Human error, or as it was formerly called, "Pilot error" is far and away the most common cause of aircraft accidents. Pilots will tell you that this is because in many cases, after the accident, the pilots are dead and therefore not able to defend themselves. Because traditional placement of controls and switches have resulted in standardized cockpits, especially in military aircraft, pilots are expected to know where they are and when to use them. Checklists are provided and because the human who is supposed to perform the tasks outlined is expected to be rational and follow the checklist, all tasks are to be completed. If a task is deleted or left out, it is then the fault of the human pilot(s) and not of the checklist writer or the human factors expert that was consulted. The verdict is again, pilot error.

Aircraft operators and crew members are increasingly being confronted with the idea that accidents are due to a breakdown of teamwork or "crew coordination." In the *bad old days* the Captain of an airliner, or the aircraft commander of a military plane was definitely in charge. What he said was the law of the air and many a co-pilot voiced the opinion that they would rather die than tell the Captain what to do. Many did just that! Now with the advent of Cockpit Resource Management (CRM) and the retirement of the old boys, flight crews are supposed to function as groups, not as individuals flying together in the same cockpit. This places an added strain on the former dictators of the air, and puts added stress on the first officers to know just when to interfere, offer suggestions, or keep their mouths shut.

Another factor in the accident equation that was supposed to take the "human out of the loop" is automation. Tasks formerly performed by the humans in the cockpit are increasingly done by computer. The human is being reduced to the role of monitor. In fact, in modern airliners, after brake release, if not hindered by airspace control authority, the pilots need not do more than program computers to make a cross-country flight, an approach, landing, and roll-out at their destination. Boredom now becomes another factor that professional aviators must deal with.

Human factors orientation

The engineering discipline has always been a large component of the practice of Human Factors. Ergonomics, or the measurement and the movement of the body relevant to the work place influences performance. Whether a control or a switch is comfortable and reachable affects the pilot's performance. It is the human body then that limits the design of the cockpit. It is important that the researcher know how the biomechanical studies are made, what are the characteristics of the subjects, and what condition they are in when the experiments are conducted.

Since many of the research projects are conducted either by the military or by universities, it could be inferred that the subjects used are young, healthy, and robust specimens. This is not altogether wrong, since, at least in their early years, the pilot population comes from these same sources - young, healthy, and robust individuals.

Research projects, by their very nature, are conducted in highly controlled, structured conditions where outside influences and distractions are fed into the equation in a scientific manner. Unfortunately, this could lead to false assumptions and often to poor conclusions. In real life, it is not often that one difficulty at a time is presented to the pilot, especially under conditions that assure continued existence. It is the compounded emergencies, experienced in rapid succession, in a terrifying environment, with death the price of a mistake that comes from and leads to human error. Add then, one more factor that draws a pilot to disaster. That is the physiological factor.

The Physiological factor

We dont often hear about physiological incidents, probably because they are not often reported. They are not often reported because many civilian pilots, simply do not know their symptoms. Oxygen deprivation, for instance, is insidious and very difficult to detect. It is almost universally associated with high altitude flight. Precious little research has been done on long period flight at the lower levels. Airline pilots flying overwater flights often endure cabin altitudes of 8,000 feet or more from 8 to 14 hours depending upon the route. Pilots complain of headaches after a flight or that they just feel bad. They blame it on the flu or some unknown disease that they "are coming down with." Smoking, drugs and alcohol, fatigue, stress, boredom, and over-the-counter medications play a significant part in physiological incidents. The effects are impaired judgment, poor memory, lack of alertness, sleepiness, and general loss of high level brain activity. It is in precisely these conditions that accidents occur. It is precisely in these conditions that experimentation pertaining to human factors must take place if we are to come any closer to solving the problems of "pilot error."

The proposal

Human or Psychological Factors cannot be separated from physiological factors when determining the correct makeup or configuration of the work place. Ergonomics alone cannot determine whether a human can interface with that cockpit environment in a stressful, oxygen deprived condition. Whether the arm measurements in healthy, robust, and physiologically healthy persons can be related to an aging, slightly corpulent, bored, partially hypoxic airline captain's arm movements is problematic. Landing gear handles are wheel shaped in modern aircraft, partially due to experiments conducted during and shortly after World War Two. Flap handles are airfoil shaped for the same reasons. Wheels up accidents continue to occur and premature flap retraction is still a problem in all flap equipped aircraft. Without the addition of physiological factors in experimentation, we may never know the optimum shape, size, configuration, or procedure that will prevent such accidents or incidents. Changing the pilot by training to suit the equipment is not the answer. It is, however, the current preferred method. It is probably thought to be less expensive to train than to change.

It is therefore the proposal of this writer that new methods of experimentation to determine "human factors" be promulgated. Real life situations cannot be realistically presented in even the best motion based, visual type simulators because they lack the basic physiological factors such as oxygen deprivation, stress, and mission importance. Even if a simulator were placed in an altitude chamber, where physiological factors could be controlled, it would be step in the right direction but still would not be quite enough. Real life motion, heat, fatigue and other factors can be simulated but the stress of mission accomplishment and the danger of death as the price of failure cannot be duplicated in current simulators.

One method of introducing realism would be to place subjects in a false cockpit in an actual aircraft, flying an actual mission, with incentives for accomplishment and punishment for failure. With current "fly-by-wire systems, the subject could actually fly the airplane while a safety pilot or pilots could electronically disconnect and take over control at any time. Perhaps the added incentive of a substantial financial award for "winning the contest" with elimination from the competition the price of failure could simulate the real life stress of "get-home-itis." Inability to accomplish the assigned tasks is a sign of failure to most professional pilots and is to be avoided at all costs.

To add further complication, separating the simulated cockpit from the real aircraft's pressurization system so that the Individuals could be subjected to hypoxia would add to the realism. The sounds, smells, motions, and frustrations would be realistic in the extreme. The scenario could be one that increases in difficulty until no pilot could ever perform the tasks, adding to the stress level. Multiple emergencies, based on actual accidents, might add another touch of realism and provide researchers with meaningful data.

Conclusions

The two disciplines, Human or Psychological Factors and Physiological Factors have been, with some exceptions, considered individually for a long time, but neither one has really approached the other in a collaborative, productive way. Psychologists take for granted that their subjects and those affected by their experimentation must be in good physical condition and devoid of physiological problems. Physiologists, on the other hand, make their physiological determinations of a subjects reactions from the basis of good physical and mental health. Neither approach is quite right. Basic research in each discipline is well along and although there is much to be learned in each arena, much more could be learned if the two could somehow get together.

Collaborative efforts are time consuming, expensive, and require specialized equipment, some of which has not yet been invented. If both psychologists and physiologists believe that accidents have causal factors rooted in each of their disciplines, then they must believe that those factors, most of the time, occur simultaneously. One can be mentally or psychologically deprived and still accomplish the mission. A pilot can be physiologically deprived and still make a safe landing. If two or more of these factors are present, what will be the outcome and what can be done to prevent disastrous results? We need to research together!

That research most probably would have to be done under the aegis of federal aviation authorities, with collaborative efforts from several aviation interests, and with funding sufficient for long term, in-depth studies. Efficient, user friendly, and above all, cockpits that support safe flight, are the goal. The time has come to change our thinking, challenge conventional knowledge, and try new things.

Part 10
PILOT COMPETENCE

37   Pilot intervention times*

P.R. Smith and J.W. Chappelow
Royal Air Force Institute of Aviation Medicine

Introduction

Few emergencies in modern aircraft require an immediate response. In multi-crew aircraft challenge and response drills and crew consultation are the norm. Even in single seat aircraft, the priority is generally to fly the aircraft and achieve a safe attitude, altitude and airspeed, before dealing with the emergency. A small number of emergencies are, however, time-critical. Total power failure in helicopters is an example of such an exception. An undue delay in reducing collective pitch demand can result in an unrecoverable decay in rotor speed. For this reason the design and certification of helicopters needs to be based on realistic estimates of pilot response times.

Reaction times have long been of experimental interest, and a great deal is known about the factors affecting speed of reaction. Some of these factors have an obvious bearing on response to total power failure in helicopters. For example: stimuli that occur only infrequently tend to elicit longer reaction times; reaction times increase as the number of possible stimuli and responses increases (in a logarithmic relationship); enhanced motivation speeds reactions; practice reduces reaction times, particularly when discrimination among

* British (c) Crown Copyright 1994/MOD: Published with the permission of the Controller of Her Britannic Majesty's Stationery Office.

alternatives is difficult; a stimulus with unique characteristics elicits faster reactions than one that shares characteristics with other possibilities; sharing attention among tasks lengthens reaction times (Boff and Lincoln, 1988). The relevance of all these factors is evident, but generalisation from laboratory to real-life tasks is inevitably in global terms rather than precise predictions. Pilots are generally well trained to identify the combination of visual, auditory and kinaesthetic cues characteristic of a total power failure, and their motivation cannot be in doubt. On the other hand, it is a relatively improbable event, and may well occur in a period of high workload or distraction. It is not possible, therefore, to specify with confidence the range of response times to be expected in real operations purely on the basis of laboratory findings.

Unfortunately, in comparison with the extensive and detailed laboratory studies of reaction times, there has been little work examining reaction times in real life situations. One unpublished trial conducted by RAF IAM in support of an accident investigation did address the issue of total power failure in a helicopter using a S61N flight simulator (Green, Chappelow and Skinner, 1991). Total reaction times ranged from 1.5 to 5.5 s, with a mean of 3.08 s and a standard deviation of 0.94 s. The aim of the present trial was to provide more detailed information on which to base recommendations regarding certification requirements for helicopters.

Method

Simulators

A Chinook Mk1 and a S61N simulator were used in the trial. The Chinook was a Redifusion six axis motion simulator with a 140° dusk/night visual system. The S61N also had a six axis motion system but the field of view of the dusk/night visual system was more limited.

Cues

A variety of visual, auditory and kinaesthetic cues could alert the pilot to the total power failure or provide diagnostic information. Video recordings were made in both simulator cockpits during exploratory flights in order to evaluate the visual and audio cues presented during a total power failure with no corrective intervention. Initial height and speed, temperature and all up weight were chosen to be representative of typical training sorties. An attempt was made at the same time to measure linear accelerations on the flight deck using accelerometers mounted between the pilots' seats on the centreline of the aircraft.

Subjects

To be able to record response times that were as realistic as possible, it was necessary for subjects to be unaware that they were participating in an experiment when they presented at the simulator. Permission to collect the data was given by those responsible for the pilots' training. Twenty-four RAF pilots were tested in the Chinook simulator, some more than once; a total of 35 exposures were recorded. Ten civilians provided response time data in the S61N.[1]

Procedure

The failure was initiated during a routine training sortie by an instructor operating a switch. No attempt was made to restrict or control the conditions surrounding the failure as we did not wish to compromise the normal training programme. The following variables were, however, noted: initial altitude; initial indicated airspeed; whether the aircraft was ascending, descending or level; whether the aircraft was turning or straight; the name of the instructor; and whether the aircraft was recovered successfully or crashed.

Data Recording

The simulators were modified so that collective position and torque could be measured and recorded in digital form at the highest sampling rate possible (30 Hz for the Chinook and 9 Hz for the S61N). The onset of total power failure was defined as the time at which torque began to decline. Responses to the emergency were considered complete when the collective reached minimum.

Detection time was defined as the period between failure onset and the first downward movement of the collective. Response time was defined as the time from the first movement of the collective to the collective reaching its minimum position. Total reaction time was defined as the detection time plus the response time.

Results

Cues

Visual Cues: The video recordings taken during the exploratory flights were analysed frame by frame for visual cues. In the Chinook rotor RPM started to decline 0.5 s after failure initiation, and reached 84% at 3.8 s after failure. At this point the generators tripped off line and the situation was unrecoverable. In the S61N rotor RPM started to decline after 0.6 s and reached 85% at 10.0s, when the situation became unrecoverable. Between these times airspeed started to reduce (at about 3.3 kt.s^{-1}), the aircraft rolled left (to a maximum of 20°) and pitched up (to 15°).

Auditory Cues: The video recording was also analysed for the auditory cues to total power failure. In both the Chinook and S61N a change in noise was detectable as soon as the emergency was initiated. In the Chinook, six unrelated frequencies in the range 310 to 5225 Hz changed by about 16% over a 2.88s period from the moment the failure occurred, and seemed to represent the forward transmission slowing down. In the S61N, the auditory cues were not as pronounced as those in the Chinook. A frequency ripple started at 7525 Hz and declined to 2112 Hz over 8.3 s.

Kinaesthetic Cues: Unfortunately, the recordings of linear accelerations made during the exploratory flights were not as useful as had been hoped due to recording problems and noise. However, it is believed that the accelerations experienced were very small, possibly less than 0.1 g, and observations on the flight deck suggested that these cues would have little if any warning or diagnostic value.

Initial Conditions

The mean conditions at failure initiation were: altitude 2724 ft (range 200 to 6000 ft) for the Chinook, 1086 ft (range 500 to 2000 ft) for the S61N; airspeed 115 kt (range 90 to 135 kt) for the Chinook, 85.7 kt (range 80 to 100 kt) for the S61N. Roughly half the cases in each simulator were from straight and level flight, and the remainder were climbing or turning.

Reaction Times

Mean detection time, response time and total reaction time for both aircraft types are summarised in Tables 1 to 3 together with the minimum and maximum values, standard deviations (SD) and 90th percentiles.

Table 1: Summary of detection time data (s)

Aircraft Type	Min	Max	Mean	SD	90th percentile
Chinook	0.37	3.62	1.13	0.65	2.01
S61N	0.66	2.53	1.65	0.60	2.52

Table 2: Summary of response time data (s)

Aircraft Type	Min	Max	Mean	SD	90th percentile
Chinook	0.54	2.93	1.15	0.56	2.17
S61N	1.21	7.37	2.58	1.82	7.02

Table 3: Summary of total reaction time data (s)

Aircraft Type	Min	Max	Mean	SD	90th percentile
Chinook	1.00	4.72	2.28	0.97	4.15
S61N	2.20	8.58	4.23	1.87	8.29

The data were log transformed to meet the assumptions of analysis of variance. Under these transforms a test of Studentised residuals showed no significant outliers in the whole data set. The transformed data also showed no deviations from a Normal distribution in terms of skewness and kurtosis. There were significant effects in the expected direction due to outcome (recovered/crashed) on response time (F=9.0833, df=1,16, p<0.01) and total reaction time (F=6.0611, df=1,16, p<0.05), but not on detection time. There was no evidence for an effect of initial conditions on detection, response or total reaction times. Helicopter type did not produce a significant effect when tested in the presence of all the other factors. However, because of missing values, this test involved a reduced data set. In addition, the initial conditions tended to be confounded with helicopter type. There is no *a priori* reason to expect a major effect due to initial conditions, so a further analysis was conducted on the factor helicopter type alone. This showed significant differences in detection time (F=4.5527, df=1,43, p<0.05), response time (F=23.41, df=1,43, p<0.001), and total reaction time (F=18.8156, df=1,43, p<0.001). The means based on this analysis are the ones contained in Tables 1, 2 and 3. In all cases the S61N times were significantly longer than the Chinook times.

Outcome

There were insufficient data on outcome (recovered/crashed) to permit further analysis in the S61N data. Of the 30 Chinook exposures for which we had outcome data, 26 pilots successfully recovered from total power failure, and four failed to recover. One of the four detected and responded to the emergency in good time, but allowed an unusual position to develop which prevented proper recovery. His data are not, therefore, relevant to the issue under consideration. The remaining three had slightly longer than average detection times and response times, but were not distinguished from the rest of the sample by initial conditions. The pattern of their responses did, however, appear qualitatively different from the vast majority of the remainder in that the collective movement was not a simple, rapid reduction in demand, but incorporated reversals or hesitations.

Discussion

It took pilots, on average, 1.13 s to detect total power failure in the Chinook simulator and 1.15 s to respond to it. S61N pilots took significantly longer both to detect and respond to the same emergency, 1.65 s and 2.58 s respectively. Mean total reaction times were 2.28 s and 4.23 s. It is clear, however, that aircraft designers and certifiers should consider not mean response times but the tail of the distribution, ie those pilots whose response times are longer than average. The 90th percentiles for the Chinook and the S61N pilots were 4.15 s and 8.29 s respectively. In fact, three of 30 emergencies in the Chinook with recorded outcomes resulted in crashes due to slow responses. Whether or not a 10% failure rate is acceptable depends to some extent on the probability of a total power failure occurring. The difference between the two samples also demands some attention.

The Chinook clearly presents a more critical problem in total power failure than the S61N, so it is, perhaps, not surprising that the Chinook pilots were significantly faster in detecting and responding to the failure. Factors other than motivation may also play a part. Chinook pilots regularly practise autorotation in the aircraft, and this probably helps to standardise the pattern of responding and encourage unequivocal reduction of the collective demand. The opportunities for such training are more limited in the commercial sector. This may account in some degree for the Chinook pilots' advantage in response time. The most significant alerting cues in both simulators appeared to be auditory. Visual cues probably provided diagnostic rather than alerting information. Kinaesthetic cues did not appear to be significant in either case. The auditory cues in the Chinook appeared to be more readily detectable than those in the S61N, and this could account for some of the Chinook pilots' advantage in detection time. On this basis three factors, motivation, training, and alerting cues, seem to distinguish the Chinook sample from the S61N sample. The first two are linked, and represent an adaptation to the shorter time available for responding on the Chinook. It seems unlikely that much could be done to increase this adaptation. The Chinook pilots' data probably represent a reasonable estimate of the best performance that can be expected in present circumstances, ie slightly more than four seconds is the minimum reaction time designers should be considering if a 10% failure rate is acceptable.

It is conceivable, however, that reaction times may be further shortened if a warning system were introduced, particularly in aircraft where the alerting cues are not very noticeable. Our results suggest there could be some advantage in attempting to mimic or enhance the auditory cues to which pilots are already sensitive. Such an ecologically valid audio warning could have several advantages: It could require less interpretation than a conventional audio warning, and could support faster diagnosis than a voice warning. It is now well recognised that the number of audio warnings a pilot can distinguish is limited; the proposed system would be in a clearly separate class, and may not need to be counted in this number.

Notes

1. We would like to thank Brintel Helicopters Ltd (Farnborough and Aberdeen) and the pilots from RAF Odiham for their co-operation and time.

References

Boff, KR and Lincoln, JE (eds). *Engineering Data Compendium Human Perception and Performance Vol III*. Harry G. Armstrong Aerospace Medical Research Laboratory, Wright-Patterson Air Force Base, Ohio, 1988.

Green, RG, Chappelow, JW, and Skinner, RJ. Response Times to Emergencies. In IAM Letter Report No 25/91, Pilot Intervention Times Preliminary Report.

38 The identification, collection and measurement of pilot competencies and flight test parameters using head-up displays

Graham J.F. Hunt, Richard Macfarlane and Jarrod Colbourn
Massey University

Introduction

The purpose of human factors development in aviation is to effect positive differences in flight crew performance. These differences have been defined in various ways but can be summarised as better pilot judgement, better crew coordination and communication, and better recognition of personal and colleagual needs, aspirations and satisfiers. While much effort has been expended in seeking to define what these construct differences might operational look like, less effort has been made in defining in operational terms the base line behaviours from which measurable statements of *better* might be made. If this observation is true, the lack of data demonstrating the transference of human factor characteristics from training and simulator environments to operational flight deck performance is even of more concern. Attitudinal measures which in some way support the usefulness of crew resource management (CRM) training in class or simulator contexts are no substitute for measures of these constructs as integral components of overall crew effectiveness. If this assertion is correct, what needs to be done?

First, identify the abilities, as compared to tasks, which describe the broad functional capabilities which contribute to individual and crew expertise. These macro abilities are the synthesis of two or more generic knowledge bases which are stored in and retrieved from long-term memory. For example, the flight crew accomplishment of *command*, defines a capacity to exercise formal, legal power and authority over aircraft crew and passengers and to establish and maintain effective and efficient crew performance.

Secondly, determine the pre-requisite attitudes, intellectual and practical skills which will provide the knowledge structures and schemata. Such schemata represent procedural knowledge (accomplishments, performances and process abilities) and the interrelationships between objects, events, and sequences of events. These pre-requisites, or *performances*, provide the particular characteristics of a given accomplishment (say, *aircraft performance management*). Such an entity, in a different constellation of performances, will provide a pre-requisite for another accomplishment (for example, *navigation management*). Competency analyses of flight crew behaviour (Crook & Hunt, 1988) have identified a hierarchy of such pre-requisites in relation to the sum of accomplishments which might describe the ability domain of flight crew performance. As an example, *command* as a macro human factor ability can be defined by six performances, each one providing a subordinate contribution to its dependent accomplishment. These performances have been identified as *captain supervising, pilot managing, managing critical incidents,* and *managing crew interactions.*

The base of the hierarchy is provided by the *specific abilities* which define each of their superordinate performances. Specific abilities may have both cognitive and affective applications. For example, from the accomplishment of *command* and its performance, *crew interacting*, cognitive abilities are included in *assessing, decision-making* and *monitoring*, and affective abilities in *leading* and *listening*

Thirdly, apply the knowledge structure to typical and atypical contexts as criteria for determining the presence or absence of acceptable *flight crew competency*. This specification of knowledge and application is called a *competency specification*. For example, in a competency specification which focuses on the accomplishment *command management*, and *pilot managing* as a critical performance, a crew resource application for this competency might be:

"Given busy radio/telephone traffic, including the issuance of amended decent profiles, the First Officer is required to brief the air crew on arrival and approach procedures in accordance with the airlines standard operating procedures. The Captain will assess the appropriateness of the plan and the alternatives which have been suggested to cope with a shortened visual approach or emergency. The Captain will decide on which strategy is best."

In this example, the performance of pilot managing is being examined through the interactive application of the abilities of such as *speaking* (oral communicating), *active listening* (being able to critically listen for relevant information) *assessing* (determining relevant environmental and interpersonal conditions), and *decision-making* (choosing the best of competing alternatives).

250

Cognitive and affective skills are embedded in the overall mastery of the accomplishment.

Finally, develop systems which can operationally measure predefined and idiosyncratic flight crew competencies.

Initial Attempts at Flight Crew Performance Measurement

Initial attempts to develop flight deck measures of flight crew competency were developed in 1990-1991 as an outcome of the Human Resource Development in Aviation (HURDA) research programme developed between the New Zealand Civil Aviation Authority and Massey University (Crook and Hunt, 1988). From the knowledge hierarchy component of a competency specification, interactive descriptors (for verbal and non-verbal behaviour between captain, co-pilot; flight crew and other information sources, eg., ATC and ground staff) were listed as a matrix for each performance and related abilities as identified in the accomplishment *command*. With this instrument, flight crew observers monitored the interactive behaviour of crews within the context of a full-flight operation. Although the data generated useful maps of flight crew behaviour which could be further analysed for profile development within phases of flight and across aircraft types, the data collection technology was clumsy.

1   The head down time associated with the recording meant that a large percentage of the processes demonstrating criterion abilities for tasks went unnoticed, and
2   The search time on the marking sheet for those criterion abilities was too long during high work load periods for the assessing officer to be confident of his or her accurate recording of those abilities.

A Head Up Display System for Data Entry

The objective for developing this technology was to allow the flight crew observer to have continuous view of the Captain and First Officer and to "snap-shot" the range of interactive behaviours being emitted as comprehensively and as immediately as possible.

Method

For the prototype device, a Reflection Technologies' Private Eye head mounted display was used. This was fixed to a David Clarke Aviation headset for testing. Assessment sheet generation was created via an IBM XT computer

251

which also contained the drive card for the Private Eye display. The positioning of the assessor in the jump-seat of an aircraft/simulator meant severe space restrictions and the lack of a reasonable surface for a conventional input device required the development of a customised input device. This consisted of a thumb operated Genius Hi point trackball placed inside a conventional Quickshot Apache I joystick handle.

The device was tested on the flightdeck of *Ansett New Zealand* BAe 146 aircraft on domestic flights for both day and night operations and in a *Hawker Pilot Trainer* (HPT) flight simulator at Massey University. For the purpose of this research the HPT flight simulator was configured as a generic light twin jet similar in performance to a Cessna Citation II.

Analysis of assessment integrity required full sortie audits from comparison with assessor evaluation. Two Panasonic WV-BL200 black and white cameras were set up for this purpose. One was positioned at the rear of the HPT to reconstruct the assessors field of view. This was set up with a Panasonic WVLAGB2 lens which shows a 55° horizontal field of view. A second camera was positioned in the front right hand side of the HPT, to provide information on non-verbal communication. This camera needed a Cosmicar HS316GX with a 90° horizontal field of view to cover the area from the First Officer's right hand to the Captain's left hand. A customised mixing unit was constructed by Massey University's Consumer Technology Department to allow the monitors and headset communication to be recorded on a single time base.

Results

Two separate trials of data gathering were set up to test the concept. In both cases three sorties were flown and assessed in real time in the flight simulator. An audiovisual recording of each sortie then allowed a panel of three assessors to re-assess each of the sorties with all the advantages of slow forward, playback, etc that videorecording machines allow. the average results of the real time assessment were then compared with the adjudication panels assessment of the sorties.

In the first trials an experienced flight crew assessor was asked to record the cockpit interactions associated with the *command* accomplishment over three sorties with different crews. A pen and paper recording system was used to identify these interactions and separate sheets were used to score them for the different flight phases which were subjectively divided into preflight, pre takeoff, takeoff, climb, cruise, descent, approach and landing, and postflight. An assessors panel was then convened to assess the same flights using the same parameters. The results of the first study are illustrated in figure 1.

252

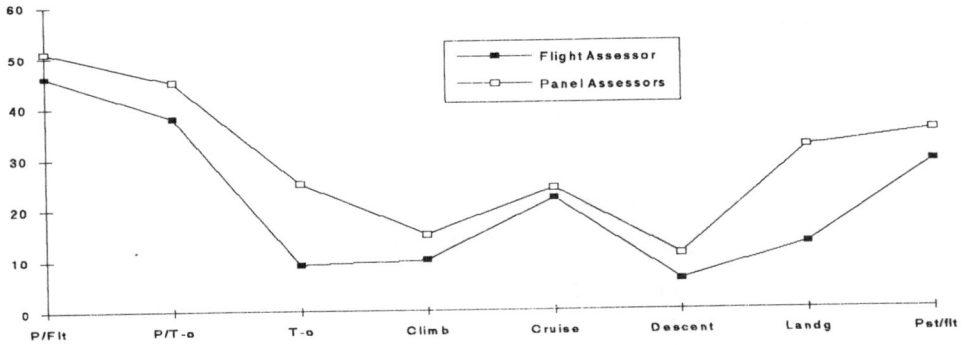

Figure 1 Effects of paper and pencil entered data

The second set of trials were run with the same requirements and parameters except that the data was entered via the headup display and input device. The results of these trials are graphed as figure 2.

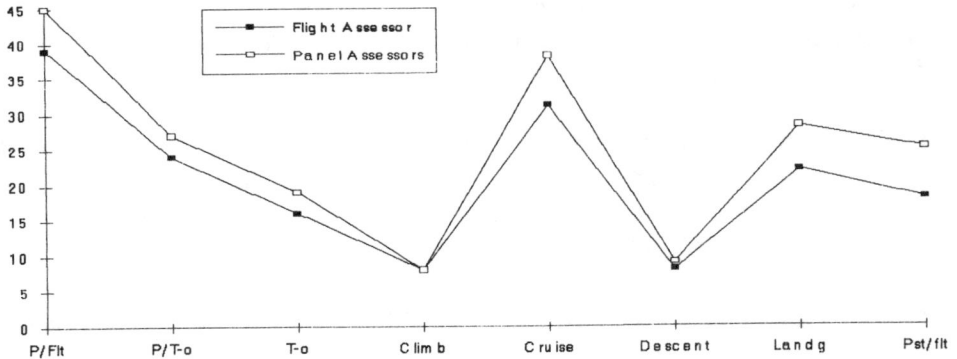

Figure 2   Effects of HUD Entered Data

In both sets of trials the panel assessors, using unlimited video play-back were able to observe more interactions than the single "real-time" flight assessor. However, the number of interactions recorded by the flight assessor through the use of the HUD device appeared to significantly improve and the difference between observed interactions of the two groups significantly reduce when the HUD technology was employed.   It is worth noting that both groups of assessors were asked to complete a questionnaire during the adjudication process to indicate the most likely reason for having missed interactions during the real time recording. The results of their responses are shown in figure 3.

As this figure demonstrates the paper and pencil data input generated substantially greater omissions such as "missed interactions while marking" than did HUD-based inputs.

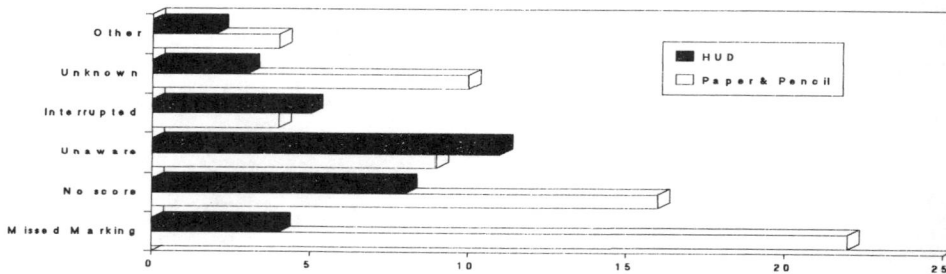

Figure 3 Reasons for Non-Identification of Interactions

Conclusions

The indications from this preliminary study suggest that the measurement of technical and human factor interactions result will be more accurate using the HUD paper and pencil recording methods. The use of the prototype device, and the reasons for doing so received unanimous support from the crews included in the pilot study. Indicative outcomes from this study suggest:

1.   Technical and human factor behaviours can be measured as integrated constructs in an operational flight deck environment.
2.   The identification, collection and measurement of pilot competencies by Flight Crew Assessors is more accurate using head-up display technology.
3.   Flight Testing Officers are more confident of their assessment using the head-up display;
4.   The head-up display (in prototype form) cannot be used when the Testing Officer is acting as PF or PNF; and
5.   The prototype head-up display is prone to glare problems in both overcast and undercast.

References

Crook, C., & Hunt, G.J.F. (1988). *Competent flight crew licensing II: the terms of reference for the Hurda Programme.* Palmerston North: Instructional Systems Programme, Massey University.

39 Complacere: an unconscious reaction to bad communication

Gunnar Fahlgren, Human Communication 193 30 Sigtuna, Sweden

(I am very well aware of the positive situation of having female pilots, so when I write 'he' I mean both 'he' and 'she'.)

The word Complacency has been used for about 20 years when explaining accidents or incidents.
All pilots are told to avoid Complacency, but nobody is told what it is and how to avoid it.

In a short questionnaire I have asked more than 1000 pilots all over the world to define the word Complacency. I found that it has quite a different meaning for different pilots. Some only had a vague idea of the meaning and were unable to describe it in words. Some used the traditional dictionary definition "over-confidence", "self-confidence" or "self-satisfaction".
Others used words like "lack of motivation", "lack of discipline", "lack of concentration" or "a feeling that others – people and/or systems – take care of all problems on board".
The answers in this survey clearly indicate that an important word, often used in our industry, has quite a different meaning for different individuals and that is not the way it should be. Important words used in important communication must be defined and clearly understood.
Therefore, my first task was to find a good definition.

Now probably all English speaking persons will stop listening. Because a great problem, which I have met, is that when I try to define or explain what the word complacency means, people do not want to know what they already think they know. For me the word complacency means a lot more than just self satisfaction, overconfidence or common laxity and sloppiness.
For me it is a Human Factors issue of great importance to flight safety.

From now on I will use the Latin word **Complacere** instead of complacency. Most people do not know what Complacere means, so by using that word I hope that I can arouse your interest and make you listen.
Complacere is a state or a phenomenon which is dangerous, if it is allowed to enter an aircraft. So first it has to be analysed and explained and then we must learn how the state of Complacere should be avoided.

My first statement, on the way to define what Complacere means, is that no pilot starts his career or will be employed in an airline suffering from Complacere from the very beginning.

The newly-appointed pilot is positive, interested and committed to his new position. But obviously Complacere can be a fact after some years.

The question is. What happens and why?

My second statement is that Complacere is unconscious behaviour caused by a change in attitudes.

As it is unconscious, the "complacered" pilot does not realise that he is in a danger zone. And then of course it is no use telling him to stop showing Complacere. To influence an attitude is much more complicated than just to tell somebody to change it.

My third statement is that Complacere might be caused by the environment we are forced to live in, the environment we select to live in, or an environment which is created by ourselves.

I would like to define Complacere as a mental state where a pilot acts, unaware of actual danger or deficiencies. He has the capacity and the knowledge to act in a competent way but – for some reason or another – this capacity is not activated. He has, so to speak, dropped his guard without knowing it.

So, Complacere is for me a state, in which a person
unconsciously does not use available knowledge.

For example,if you cross the street on a red light but look around, that is not Complacere. You might be breaking the law but you are not "complacered" because you are aware of the safety aspects, and you use that knowledge in a constructive way.

On the other hand, if you cross that street on a green light, but do not look around before you start your run, you are "complacered". You know the danger of being killed by a careless driver, but you do not activate your knowledge in this specific situation. This does not mean that you suffer from general Complacere, but I would say that you suffer from Traffic Complacere and that can be fatal!

This indicates that there are different kinds of Complacere. The word Complacency is just a word indicating a general state. Complacere, on the other hand, is a Human Factors issue of great importance to flight safety. So to fully understand, what it is all about, I will select three different varieties of Complacere, which are specially dangerous in a cockpit.

Technology Complacere, Leadership Complacere and Management Complacere

Complacere can be developed in any environment where a person feels his skill, his knowledge and experience are called to question by superiors.

It can strike a person in any occupation. However, a pilot might run a higher risk of being "complacered" due to the adopted seniority system, which hampers the possibilities of leaving the environment and going to another airline. He is caught in a trap, unable to change his working situation. His attitudes might change and Complacere can be a result of gradually hampered creativity.

When we talk about attitudes we should remember that an attitude contains three important elements.

1. * The affective part *   our **feelings** for the object.
2. * The cognitive part *   our **knowledge** of the object.
3. * The conative part *   our **behaviour**.

Very often our behaviour is a result of our feelings and our knowledge. Then there is a "balance" in the system. But sometimes our behaviour does not correspond to our feelings and knowledge. That state is called Compliance. Compliance is so to speak a false attitude, valid when behaviour does not correspond to feelings and knowledge. A compliant situation can arise e.g. when we very much want to please a group or to please a person.
By the way the word please stems from the Latin word Complacere.
Also the word Compliance is well known to English speaking people and can give a false experience of its meaning to others than psychologists.
Medical doctors like compliant patients.
For an airline captain a compliant first officer can be a disaster.
So, from now on I will use the Latin word **Complaceo** instead of compliance.
There is not only a linguistic connection between Complaceo and Complacere.
In my opinion there is also a psychological connection, which I will try to prove.

Technology Complacere

It is very easy, in our highly technical world, that we fall into Technology Complacere. We often feel like being forced into uncritical belief in technical authority. We gradually get a feeling that technical systems take care of all problems on board. The knowledge and feelings that a technical system very well can fail are pressed further and further into the background. In that way technology can influence us to a change in behaviour and once again we can talk about Complaceo. This Complaceo might lead to Complacere. In that way active knowledge and skill is set aside by an uncritical belief in technical systems.
That is Technology Complacere.

Some examples of Technology Complacere.
1. The captain on the TITANIC probably felt forced to behave as if his ship was unsinkable. Although his knowledge that icebergs and ships at high speed can cause a catastrophic situation he cruised at maximum speed in dangerous waters. He trusted technology and behaved contrary to his knowledge and experience.

2. "Complacered" stockbrokers on Wall Street so trusted technology that they let the buying and selling be controlled by computers only, with no input from themselves, and the result was a crash in October 1987.

3. A technically "complacered" pilot can set his autopilot in altitude hold at 2000 feet. When the speed increases, due to descent, his only reaction is to reduce power and finally he ends up in the Everglades National park.

4. Let us look at the Stockholm accident, December 1991, with the MD:80, which successfully landed in the woods north of Arlanda International Airport after engine failure on both engines.

257

One detail in this accident is an example of Technology Complacere. Both pilots definitely knew that power has to be reduced on all kinds of engines (Cars, Aeroplanes, Lawn- Mowers) if they start to bang. In this case they so trusted technology and the auto throttle, that they unconsciously believed that it would take care of the problem. It did not.

The auto throttle was not programmed for this kind of incident and increased power, which caused the engines to disintegrate.

5. A pilot riding a bicycle knows that he has to increase power on the pedals if he loses his tailwind and still wants to keep up the speed . He also knows that the speed indicator in an aircraft, which actually indicates pressure and not speed, will give a higher value when meeting headwind in a windshear and that head- wind demands more power in order to stay stabilised in an approach. So, if a pilot believes that he should always reduce power when speed increases, that is Technology Complacere.

Because in a tailwind approach, when the tailwind disappears or is reduced clos- er to the ground, he must increase power when airspeed increases.

Here we can find a reason for Controlled Flights Into Terrain, during approach, due to a trained reaction always to reduce power when speed increases.

6. All pilots know that the pressure altimeter indicates pressure and not altitude. We also know that strong winds over mountains cause low pressure. The altitude limits, presented in the landing chart, give 1000 feet clearance above mountains.

If a pilot, in mountainous terrain and strong winds, flies at the altitudes given in his landing chart and regards the instrument as an altitude indicator, which gives him 1000 feet separation, it is Complacere.

He has lost margins without knowing it.

Many accidents and Ground Proximity Warnings are caused by that type of Complacere. To fully trust a Flight Management System in such terrain is also Technology Complacere.

If not corrected the autopilot will then fly at a too low altitude and very close to disaster and an increase of Controlled Flight Into Terrain accidents due to this is already a reality.

7. All pilots also know that an EPR instrument only indicates a difference in pressure between the front and the rear of a jet engine. If he puts that knowledge aside and acts as if that EPR-instrument is an indication of power, it is Technology Complacere. Then his flight might end up in the Patomac river.

8. Let us say that you are going to make a take off with a DC-9. You have ap- plied all necessary corrections and max. weight calculation gives 49 tons.

But the centre of gravity, during take off, is in an extreme forward, but accept- able, position. All pilots know that in this case the stabiliser is in a very negative position to press the tail down in order to compensate for the heavy nose.

If you then believe you are likely to be safe, independent of your centre of gravi- ty, you are Complacered and are off your guard without knowing it.

Namely, when airborne the load on the wings is aerodynamically much higher, then indicated in the Load Sheet, with a nose heavy aircraft compared to a less nose heavy aircraft.But you are probably not arguing against the technically con structed Load Sheet. That is Technology Complacere.

258

Leadership Complacere

This type of Complacere is mainly a leadership syndrome caused by bad communication by the leader himself. A captain, unaware of his behaviour, can create an atmosphere where his crew members feel tense and uneasy.

Maybe the captain does not listen, maybe he is irritated, and has a negative boddylanguage, which will cause a very negative atmosphere in his cockpit.

In this tense atmosphere his first officers will soon stop providing him with their support, their knowledge and experience and even withhold their doubts regarding, what the captain does. A wrong action might not be corrected.

It might even go so far that the first officer is waiting, with pleasure, for the captain to make a mistake.

In this case necessary feedback on the captain´s performance is killed by bad relations. The captain has, so to speak, dropped his guard without knowing it due to bad relations on board, which is Leadership Complacere.

Some examples of Leadership Complacere.

Very experienced pilots have for many years been flying as captains in smaller airlines in Africa, the Middle East and Far East. When the native pilots, after some years, have got enough flying hours and are qualified for an ATPL, they become captains and the senior, non native, captain is set in the right seat as first officer. Usually there are no safety problems related to this fact, but in certain cultures the "Captain" is not mature enough to take "criticism" from a first officer. The experienced pilot in the right seat is very well aware of that and also knows that if he interferes in the captain´s flying and decisions he will probably be fired and sent home, even if his interference was both right and valuable with regard to safety.

The first officer´s Complaceo will result in Complacere for his captain.

In some cultures, where a captain´s decision cannot be questioned, the same condition might be the result with a native captain and first officer in the same cockpit.

Many years ago a BAC 111, took off from London with a captain and first officer on board who did not cooperate at all. On this specific take off the captain ordered "Slats in" or said just "in" pointing at some control. The first officer took it as an order to take slats in even if it was much too early.

The aircraft stalled immediately at low altitude and the result was a fatal Leadership Complacere accident.

In these examples the captain is suffering from Complacere due to his attitude and he will lose a lot of information without knowing it.

Leadership Complacere is a fact.

The difference in accident rate between regions might be caused by Leadership Complacere.

Management Complacere

Management Complacere is very similar to Leadership Complacere, but instead of bad communication between individuals, Management Complacere will develop in an environment with bad communication between an individual and the system in which he is working.

Management Complacere might e. g. develop in a situation where communication is bad between **management** and pilots. In terms of Transactional Analysis this communication might be experienced as PARENT to CHILD messages.

The pilot feels criticised, controlled and reproved by the the management.

The pilot´s active knowledge, creativity and motivation is gradually killed by bad communication. He still has the knowledge but is not stimulated to use it. Gradually he might develop a negative attitude to his company and to his work.

As nobody is asking for his knowledge and feelings regarding Flight Safety, or other important issues, he might gradually be transformed into a person who does not give information any more. He might even stop asking for information.

Then he won´t ask his colleagues and finally he even stops asking the own memory for knowledge. His behaviour can come into a Complaceo state and this Complaceo opens the door to Complacere. Thus Management Complacere has been developed and the pilot is unconsciously not using available knowledge.

Management Complacere is a danger to Flight Safety and has to be compensated for by the other pilot in the crew.

If both pilots suffer from it, then an accident can be near by.

Now I can hear your question: "How can Complacere be avoided?"
My answer and my keyword to that question is *A Feeling of Importance*

All leaders must think of the importance of creating such a climate in the company that all employees have the feeling of being important.
They have to be listen to and they must be respected.
That will prevent Management Complacere.

Any person who is set to operate a complicated, automatic technical system, must be given a clear and strong feeling of being important and a feeling and knowledge that he is expected to interfere and correct the system when necessary. During training this interference also has to be trained.
That will prevent Technology Complacere.

In smaller groups it is of a great importance that the leader creates an atmosphere in which all group members have a feeling of being important and listened to.
That will prevent Leadership Complacere.

Complacere can strike any pilot.
Complacere can be at hand early in a pilot´s career and it can strike the pilot many years later.
The real "stick to the rules" pilot can be complacered as well as the "computer-freak" pilot.
We are all in the danger zone, so keep your guard high and fight against Complacere.

40 Pilot's psychology through an oculometric study

Eddy L. Racca, Aeroformation

General

Aeroformation has implemented a study on the visual pattern of pilots, using oculometric methodology, in order to detect any difference in the research of information between a "semi-classical cockpit" A310 and a "full glass cockpit" A320. We think that we have through it obtained very interesting fallouts in the area of pilot's behaviour, cognitive and visual strategies.

First of all, it is important to note that the average of the results is statistically significant (as shown by controls with mathematical tools like W of Kendal) and strongly suggests that all the pilots have globally the same visual strategy, whatever is the type of aircraft, and their position in the cockpit, pilot flying or pilot non flying. Anyway, around this general behaviour, we can observe individual strategies, related to different parameters and this will be now the subject of our presentation.

Experimental device

For this study, we worked in Full Flight Simulators, with outside night version, in A310 and A320. The two pilots were equipped with oculometric helmets, NAC 5 type, as shown in figures 1A and 1B, where we can also see the pilot in charge to simulate ATC traffic, and to manage the simulator ; on the left of figure 1B the two screens allowing to monitor the quality of the records are visible.

The subjects were 6 simulator instructors and 4 instructors and check pilots from Aeroformation and Airbus Industrie, as it was our first contact with this methodology, and that we wished to fine tune it with our own personnel ; at the end of our experimentation we have been able to work with an Air France Captain and his results have confirmed those obtained with our pilots.

Methodology

To avoid to the pilots to wear the helmet a too long time, which is difficult to do, we made all our records through ILS or VOR-DME approaches, so that the duration of each was around 10 to 15 minutes.

After each session, we showed the video recording to the pilots and they were asked to comment the visual pattern, the variations of the eye position, as shown by the oculometric device. This was necessary to avoid any misinterpretation related to a shift of the initial calibration, and to obtain a feedback from the pilot. This part of the experimentation has been very useful for us on the psychological point of view. We have observed that the visual pattern of a pilot is apparently not the result of a conscious decision, but as soon as he can see it on a screen, he better understands his way of functioning, himself he analyses it thoroughly, and finally he is able to change some "bad" habits and to improve himself.

All the pilots have been very happy to "see" their scanning and this will conduct us to try to implement a pedagogical tool to help the pilots in this area, mainly those coming from old gauges classical cockpits or having some difficulties to the use of the CRTs, EFISs and so on.

Detailed results

Presentation of the records
All the results will be presented here under the form of chronograms, where the time (in centiseconds) is on the horizontal axis, and the different scanned zones (as defined on the different screens) are on the vertical axis. To avoid a confusing display, we have not given the definition of each zone, we will provide the name of each when necessary for the comprehension of the figure.

One airplane, one approach, five pilots
The first part of this presentation will be related to the A320, to an ILS approach, with five different pilots, all seated on the left hand seat and acting as pilot flying. Obviously, they all have a common "anchoring point", that means the zone around which they organize their visual pattern, and it is A, the Flight Directors bars ; this is very interesting because they are a synthetic representation of a lot of information, so it is extremely important to use them to have the "big picture" of the situation.

Otherwise, having a quick look to figures 2 to 6, it is obvious that around this common basis they develop different visual strategies.

Fig. 2 - Former military pilot on single seaters, then a short experience on general aviation, now for a very long time simulator instructor, flying occasionally on a light aircraft. He has a very expansive strategy with frequent changes from one point to another ; secondary anchorings are made on G (ILS indications), on I (Flight Monitoring Annunciator) and a little less on M (Flight Control Unit) ; this is normal because as an instructor in charge of the standardisation of courses, he insists very often on the necessity to use and crosscheck FMA and FCU. In the end of the approach, he his more closely monitoring the vertical speed (E) than at the beginning and he does not use at all the peripheral vision (Z, not on fig. 2).

Fig. 3 - Former navy pilot on heavy patrol aircraft, now simulator instructor but for a shorter time than the previous, and chief pilot of an aeroclub, so flying frequently. His strategy is very economical, the secondary anchoring is on G (ILS indications), and there is a light anchoring in L (Navigation Display).

Fig. 4 - Former military pilot with a strong experience on tankers, now simulator instructor for a few years. He has a very particular strategy, anchoring on the first half of approach on I (FMA) and less on M (FCU), then on the second half anchoring on A (FD bars) and G (ILS indications), then F (Radio Altitude). This strategy is rather economical in the first half, less in the second.

Fig. 5 - This subject is a test pilot, former military fighter, but having acted as an airline pilot for 9 years before joining Airbus. His ocular strategy is typical of an "expert" one, very economical. We must not be confused by the frequent movements from A (FD bars) to I (FMA) because A is in the middle of the Primary Flight Display, and I is in the upper part, the distance between them beeing less than 3.5 inches.

We observe also the use of peripheral vision (Z) that is very economical to have a general view of a situation.

We have seen this strategy on most of the pilots.

Fig. 6 - This subject is an instructor pilot, former military fighter pilot in combat units of very high level, and he has kept from this period the habit to have accurate looks on several points. So, he uses very often the peripheral vision (Z), the altitude (D), the outside (O) ; in summary he is the exception among the pilots.

A Crew

Now, we study a crew made of two test pilots, the pilot flying being the one of fig. 5, the pilot non flying having been for ten years a navy pilot on heavy patrol aircraft, then for ten years on airline pilot ; they are both flying an A310 in a VOR-DME approach.

We observe a very good complementarity and task sharing between the two pilots ; the PF is strongly anchored on A (FD bars), monitors the flight parameters (C, F, M, U) and has a strong use of the peripheral vision, and finally his ocular scanning is rather expansive.

The PNF has a very economical strategy, and uses moderately the peripheral vision ; he cross-checks some parameters with the PF (A, F, M) and monitors some others more specifically (B, T).

Independently of the work of the crew, we can observe that the PF has a more expansive strategy on this VOR-DME approach on A310 than he had on an ILS approach on A320 (fig.5).

Permutation of roles

On figures 9 to 12, we have two pilots exchanging their roles of PF and PNF during two successives VOR-DME approaches on an A310.

Pilot X has been a military pilot, mainly on fighters or light bombers.

Pilot Y has always been a civil pilot, flight instructor on twin propjet aircrafts. Both are now simulator instructors.

First, it is obvious that pilot X has a less economical visual strategy than pilot Y, whatever are their respective positions.

Pilot X had the habit to be on a single seater so, when he acts as a PNF, he has a strong tendency to closely monitor what the PF is doing and his strategy is more expansive as PNF than as PF.

Pilot Y has been instructor for a long time, he is used to the role of non directly piloting crew member, and so at the opposite of pilot X his strategy is more economical when he acts as a PNF than when he is a PF.

Anyway, we can observe that again there is a very good task sharing and complementarity between the two pilots.

One pilot, two airplanes, two types of approach

We come back now to the pilot of figure 2, and on figures 13 to 16 we see him acting as a PF in different situations.

Generally speaking, he kepts his expansive strategy, "beaming" from an anchoring point A (FD bars), just a little bit less used in A320 VOR-DME approach.

Except this common characteristic, he uses secondary anchor points, more or less, related to different types of aircraft or approach :

- G (ILS indications) when flying an ILS, A310 or A320.
- M (Flight Control Unit) is less used in A310, ILS.
- L (Navigation Display) is mainly used in VOR-DME, A320.

- D (Altitude) is mainly used in A310, because on this aircraft, the altitude is not presented on the Primary FLight Display but on a separate classical instrument.

One pilot, one airplane, different roles and approaches

Finally, on figures 17 to 20, we come back to pilot X (fig. 9 and 12) flying now on A310, as PF or PNF, in ILS on VOR-DME approaches.

For him, whatever are the aircraft and his position on board, his anchoring point is mainly A (FD bars).

His strategy is not very economical ; interestingly enough he uses widely the peripheral vision, excepted when he acts as a PNF in an ILS on A320, having at this time his most economical strategy and spending a long time reading his documentation (T).

Conclusions

So, we have seen that around a common behaviour, each pilot has his ocular strategy, related to the aircraft, to the type of approach and to his role in the cockpit.

Through the interviews we have conducted with the participants to this experiment, we have understood that even if this behaviour is not fully conscious, as soon as the pilots see their scanning on a screen, they are able to immediately explain why they do so, and they have a strong memory of what they have looked at, during the approach.

It has been also obvious for us that the presentation of information on CRTs changes the scanning, compared to a classical cockpit with different separated gauges on instruments, but that allows the pilot to have a more economical ocular strategy ; with CRTs and with synthetic presentation like the Flight Director, when a situation has been well established, the use of the peripheral vision allows to monitor the flight parameters with a minimum of ocular expense.

Finally, to help pilots having difficulties with the use of CRTs, for any reason, we intend to develop a pedagogical tool, that could be a video showing the visual pattern of an "expert" pilot during different types of approach, with the pertinent comments of this pilot.

We also intend to continue this study, with an other type of oculometric helmet, more easy to wear and more reliable on the stability of calibration, with airline pilots of different ages and qualifications, and in flight, but in this case for obvious safety reasons, only one pilot will be equipped with the helmet.

Figure
1A

Figure
1B

The Airbus Training Organization

OCULOMETRIE02/9400

266

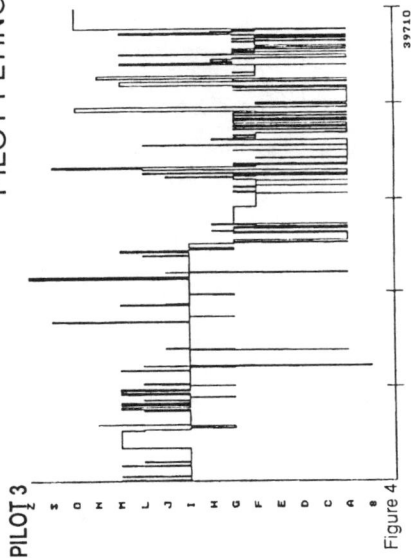

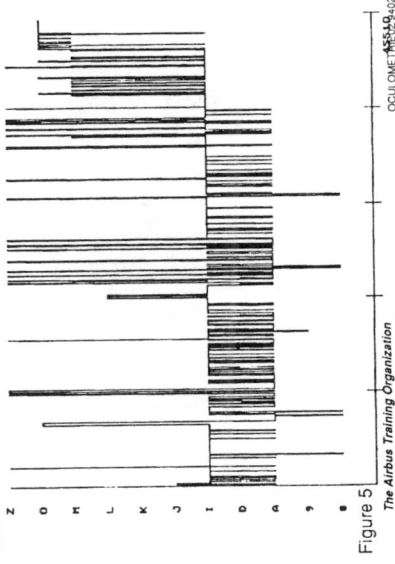

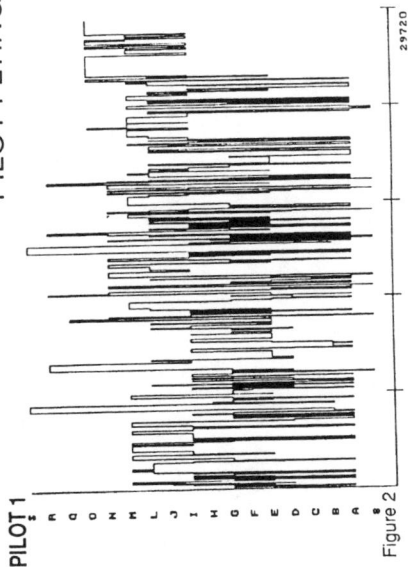

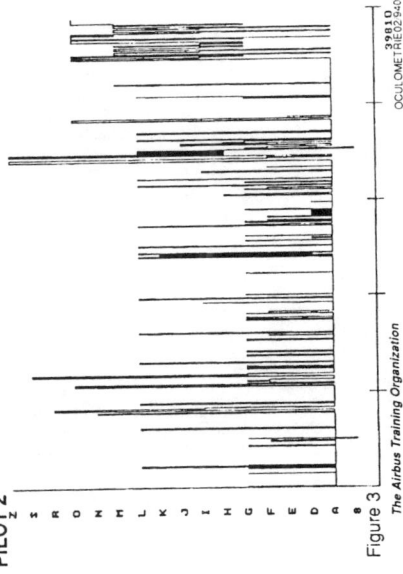

267

ILS A320

PILOT FLYING

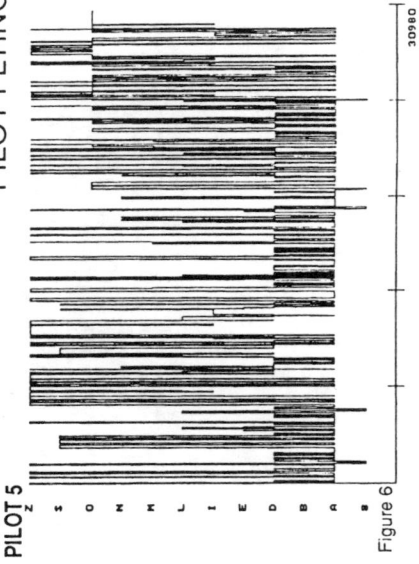

Figure 6

OCULOMETRIE 02 9-03

VOR A310

CREW

PILOT FLYING

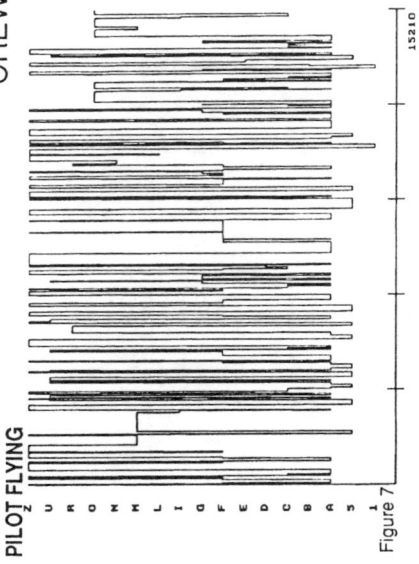

Figure 7

PILOT NON FLYING

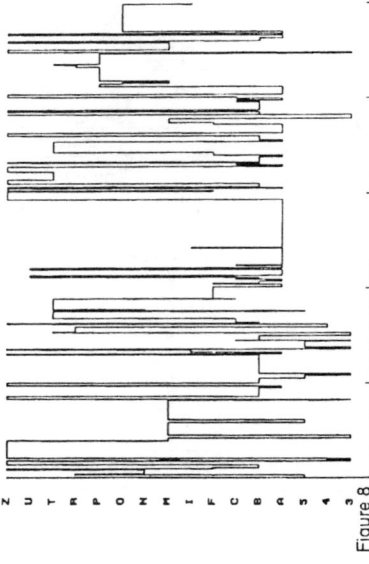

Figure 8

OCULOMETRIE 02 9-04

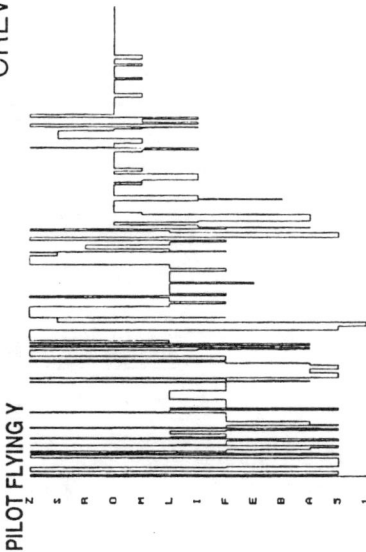

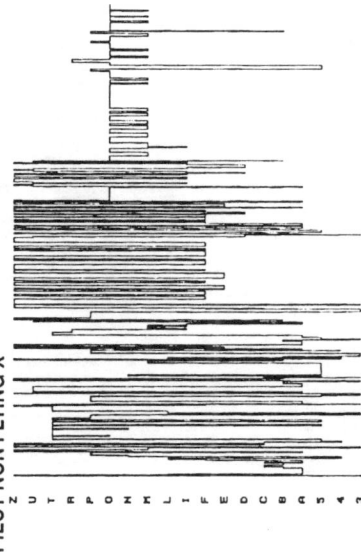

VOR A310

CREW

Aeroformation

PILOT FLYING X

Figure 9

PILOT FLYING Y

Figure 11

PILOT NON FLYING Y

Figure 10

The Airbus Training Organization

OCULOMETRIE 02 9405

PILOT NON FLYING X

Figure 12

The Airbus Training Organization

OCULOMETRIE 02 9406

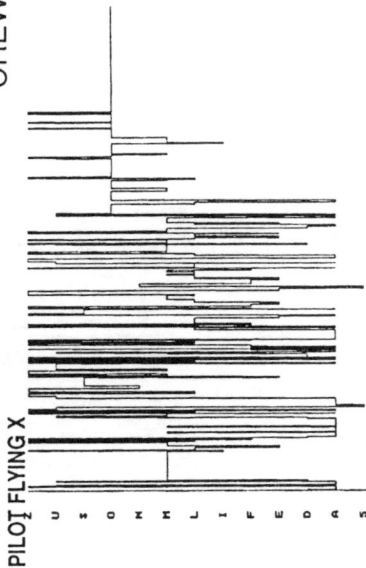

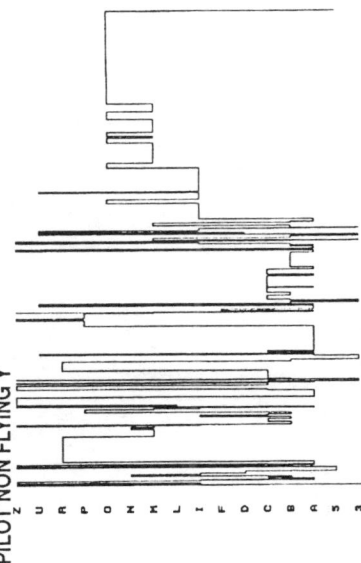

269

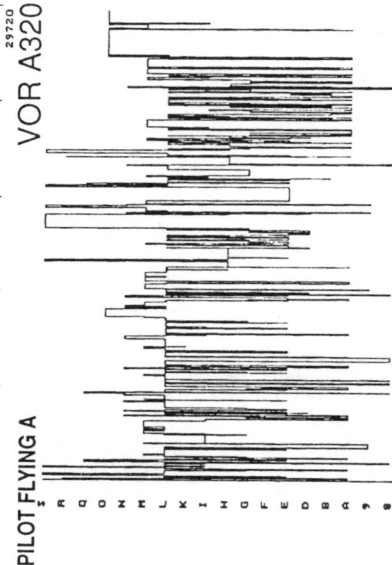

PILOT FLYING A

Figure 15

VOR A320

29720

OCULOMETRIE 02.9408

PILOT FLYING A

The Airbus Training Organization

Figure 16

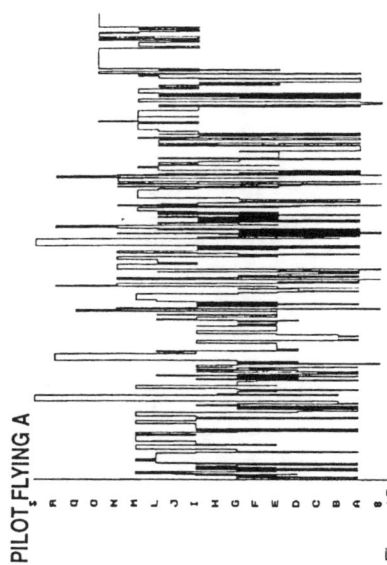

PILOT FLYING A

Figure 13

VOR A310

30300

OCULOMETRIE 02.9407

33810

PILOT FLYING A

The Airbus Training Organization

Figure 14

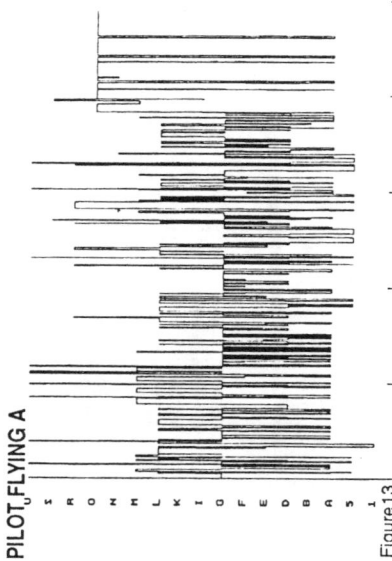

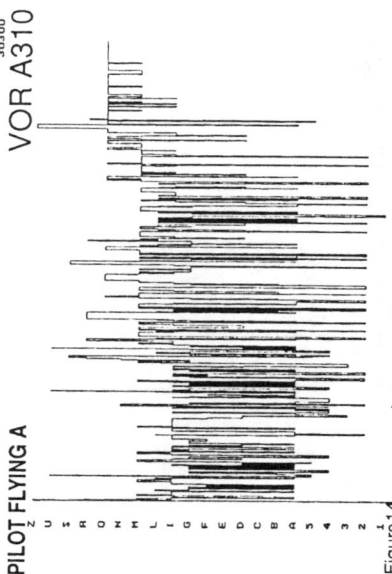

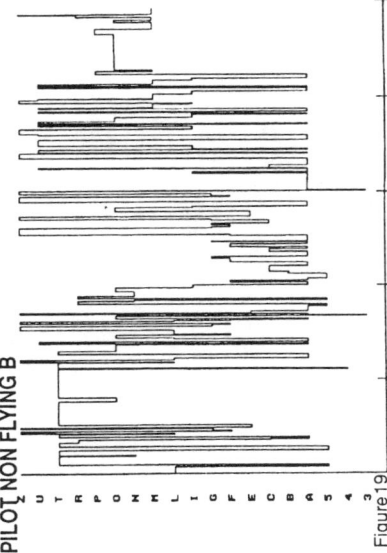

ILS A310

PILOT FLYING B

Figure 17

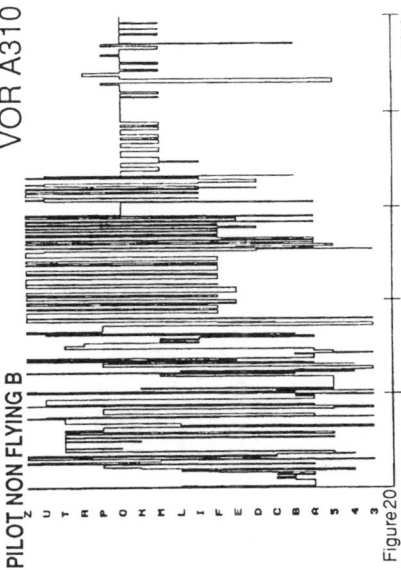

ILS A310

PILOT NON FLYING B

Figure 19

OCULOMETRIE 02 94 10

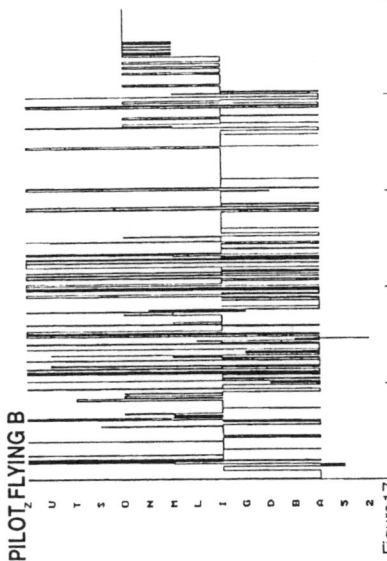

VOR A310

PILOT FLYING B

Figure 18

The Airbus Training Organization

OCULOMETRIE 02 94 09

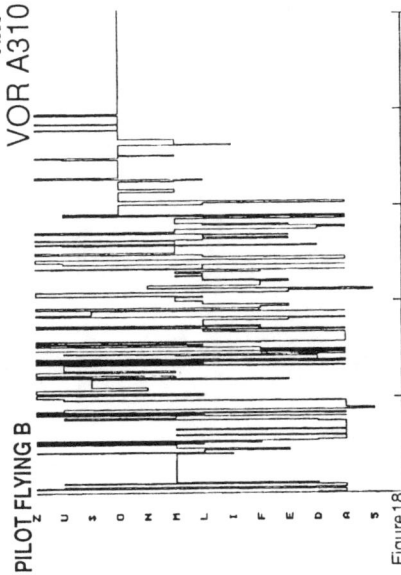

VOR A310

PILOT NON FLYING B

Figure 20

The Airbus Training Organization

OCULOMETRIE 02 94 10

Part 11
SITUATION AWARENESS

41 Measurement of situational awareness and performance: a unitary SART index predicts performance on a simulated ATC task

R.M. Taylor, S.J. Selcon and A.D. Swinden
RAF Institute of Aviation Medicine, Farnborough, UK

Introduction

Loss of "situational awareness" (SA) limits operational effectiveness and has been a significant factor in recent aircraft accidents. SA refers to the pilot's knowledge of the tactical environment. Theories stress the importance of the pilot's continuously updated, cognitive model of the situation, influenced by attention and working memory limitations, and by the availability of knowledge of the important relationships, stored as schema and scripts. Problems with SA seem to be associated with the increasing complexity and automation of aircraft cockpit systems, and the effects of high workload and poor cockpit information. Military pilots complain of being "swamped with data and starved of information". But, automation-induced complacency on highly automated civil flight decks can lead to similar risks of inattention and loss of SA. Nevertheless, SA seems particularly important in highly dynamic environments requiring fast anticipatory responses. In emergencies, expert decision makers first classify situations in terms of recognised categories of experience, before selecting decision actions, i.e. "recognition-primed decision making". In the military, accurate assessment of a changing situation is the basis for effective reactive planning. During air combat, for instance, SA determines "who shoots" and "who chutes". Thus, improved SA is the driver behind developments in aircraft helmet-mounted displays and large area cockpit displays.

The ability to measure, model and predict SA is needed for systematic SA improvement, either by training, or by design. But SA is an unobservable cognitive state, not directly available for analysis. Thus, measurement presents

275

practical and theoretical difficulties, affecting the validity and reliability of the measurement data. Techniques include performance metrics, physiological indices, and memory probe measures of knowledge. Following the example of workload metrics, we have sought to develop a subjective rating technique for estimating SA for easy implementation in the field, when performance and other intrusive measures are impractical. SART (Situational Awareness Rating Technique) is based on aircrew constructs for SA elicited using the Repertory Grid method (Taylor, 1990). SART provides multi-dimensional SA rating scales. There are three primary SART rating dimensions, namely Demand and Supply of attentional resources, and Understanding (3-D SART). 10 secondary rating dimensions are nested within the primary domains (10-D SART). Validation studies have indicated sensitivity, and predictive and diagnostic ability, on a range of tasks (Selcon & Taylor, 1990; Taylor & Selcon, 1990). But evidence is limited for performance in continuously demanding operations.

A unitary measure of subjective SA could have practical utility where a clear recommendation is required on the best system interface or task design for SA. Simple ratings of SA have shown limited sensitivity. We have proposed a method for deriving a unitary estimate of SA from SART ratings, which retains the characteristics of validity, sensitivity, and diagnostic power embodied in the individual SART dimensions. The unitary index is obtained by combining the rating means on the three principle SART dimensions, or the equivalent rating means from the 10-D SART, using the following algorithm:

SA (Calculated) = Understanding - (Demand - Supply);
or SA (c) = $\sum U/N_u - (\sum D/N_u - \sum S/N_u)$.

The formula was derived from *a priori* theoretical considerations of how the constructs interact, rather than empirical or statistical evaluation. Hence, there is some arbitrariness in the choice of formula that necessitates caution in the weight that should be attributed to it. *Post hoc* validation comparing the calculated SA measure with experimental results has provided mixed evidence for the utility of this approach. In a study of bi-modal cockpit warnings, SA(c) estimates from 3-D SART were found to reflect the performance to some extent, although not fully (Selcon, et al.,1992). In a flight simulator study of tactical operations, SA(c) estimates calculated from 10-D SART ratings were sensitive to experimental conditions and reflected performance data. The 10-D SA(c) was more sensitive than memory probes, and judged as possibly the preferred single metric (Crabtree et al., 1993; Vidulich et al, 1993).

The present experiment was conducted to provide a more stringent *a priori* validation of the 3-D SA(c) estimate, and to provide comparisons of 3-D and 10-D SA(c) estimates. The experiment used a computer-based Air Traffic

Control (ATC) Simulation task developed by DCIEM Canada. The *a priori* prediction was that ATC performance improves with increasing SA(c).

Experimental Method

Twelve non-aircrew subjects participated in the experiment. All subjects were staff at RAF IAM, between 19 and 30 years of age. Subjects were naive to both the simulated and real world ATC tasks. All were regular computer game players. All received the same amount of practice on the task. Training was provided until a successful performance criterion was reached. Subjects were presented with three experimental scenarios, containing 3,6, or 9 aircraft for control, selected to provide a range of task demand. Each scenario was of 5 minutes duration. The order in which the scenarios were presented was balanced to prevent order/sequence effects. Aircraft entered the display screen from predetermined headings and positions, at fixed "irregular" intervals, distributed to give a gradual build up of aircraft on the screen early in the scenario. The task was to control the aircraft to exit the area safely, avoiding conflicts, along exit path headings and at altitudes shown on the screen schedule. Each scenario was paused twice, at equally spaced time intervals, for 3-D SART and simple SA(r) ratings. At the end of the scenario, 3-D SART, 10-D SART, and SA(r) ratings were obtained. Performance was scored by the number of aircraft controlled correctly, leaving the area at the required altitude and direction, expressed as a percentage, and by the number of conflicts, comprising collisions and near misses.

Results

Analysis of the performance data and of the ratings was by ANOVA across the 3 scenario conditions, and by correlation of the individual measures. Learning and order effects were small. ANOVAs showed significant effects on the performance measures of %Correct (F= 246.79, df 2,22 ; p< 0.001) and Conflicts (F= 8.75, df 2,22; p<0.01). Post hoc tests showed the 3 aircraft condition to have more %Correct scores (p<0.001) and fewer Conflicts (p<0.005) than the other two conditions, but there were no significant differences on these measures between the 6 and 9 aircraft conditions. This strong conditions effect was repeated in the ratings data, including the SA(r) ratings, the individual 3-D SART ratings, and the SA(c) estimates calculated from both the 3-D and 10-D SART data (p<0.01 or better). Only three of the 10-D SART ratings (Information Quantity, Information Quality, Familiarity) showed no conditions effect. Significant differences between ratings for the 6 and 9 aircraft conditions were obtained for 3-D Demand ($p<0.01$), and for 10-D Variability ($p<0.01$) and Spare Mental Capacity ($p<0.05$). The correlations in Table 1 summarise the main findings. Correlations greater than 0.53 are significant at the 5% level. The SA(c) estimates showed no reduction in prediction of performance compared with the simple SA(r) ratings. Better general predictions of performance were obtained from the 3-D SA(c) and SA(r) ratings for %Correct than for Conflicts. 3-D SA(c) provided the best prediction of %Correct performance, accounting for 40% of the variance in the data. However, 10-D SA(c) improved the prediction of Conflicts. Figures 1 and 2 show the relationships between 3-D SA(c) and performance.

Table 1 Correlations between dependent variables

	% CORRECT	CONFLICTS
SA (r)	0.581	-0.346
SA (c) 3-D	0.640	-0.468
SA (c) 10-D	0.526	-0.548

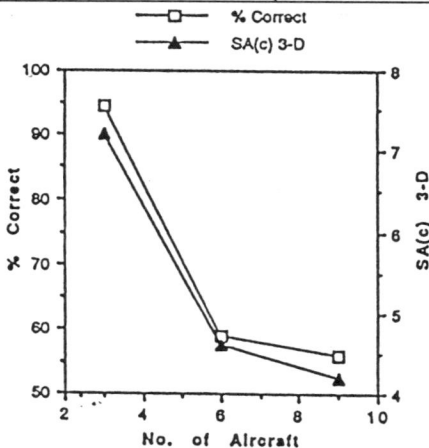

Figure 1 3-D SA(c) and %Correct

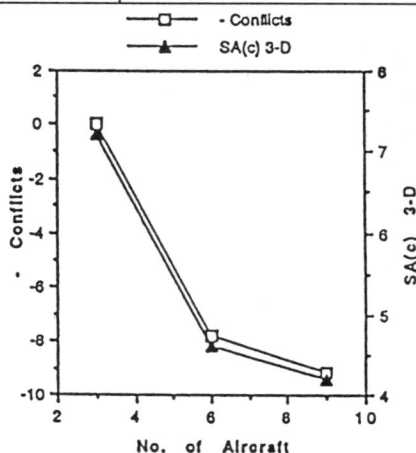

Figure 2 3-D SA(c) and Conflicts

Discussion

The experiment provides evidence for the validity of *a priori* predictions of task performance, based on a unitary calculated measure of SA. Since the SA(c) formula was based on earlier findings, and not merely a best fit to the current data, the test is a more stringent assessment of the validity of the measure than the previous *post hoc* analysis. The individual SART dimensions exhibited differential sensitivity to the experimental conditions, and showed differences between the 6 and 9 aircraft conditions not exhibited in the performance data, nor in the SA(c) and SA(r) scores. Thus, the multiple ratings contributing to SA(c) can provide sensitivity and diagnostic information when performance measures are unaffected.

SART was not developed to provide a unitary measure of SA. The initial aircrew knowledge elicitation presented SA as a multi-dimensional construct. The validity of using the combined SA(c) estimates needs careful evaluation. By combining the SART scores, the assumption is made that SA can be usefully represented as a uni-dimensional concept. This is because the single SA(c) estimates represent a scalar as opposed to the 3-D vector quantities of the separate 3-D SART scores. The outcome is analogous to deriving a single measure of colour discrimination ability when colour vision is most usefully considered in terms of the ability to discriminate differences in 3-D colour space. Arguably, the required metric of SA needs to discriminate differences between situations that are important for decision making effectiveness, rather than to provide a simple quantitative index of SA.

The SA(c) formula derives from the proposition that SA is principally concerned with knowledge of the important relationships, and of the status, of variables in the situation. It is considered that SART Understanding ratings reflect knowledge of important relationships between situation variables, which largely determines SA. The ratings of SART Demand and Supply indicate the matching of attentional resources to changes in the situation variables. This attentional matching provides information on the current status of the variables. It acts as a modifier of SA, independent of knowledge of the important relationships, providing refinement and updating of the situation model in accordance with the changing status of the variables. Attentional matching increases SA when the available resources are sufficient (S>D), and reduces SA when the resources are insufficient (D>S).

This formulation is highly simplistic in contrast with complex multi-variable information processing models of human cognition, but it merely seeks to provide the best estimate of SA from the three SART scales. The simple mathematical treatment of the means of the SART ratings contrasts with the complex conjoint analysis procedures used to combine workload rating dimensions (e.g.SWAT), and the weighting of rating dimensions (e.g. NASA TLX). The psychometric properties of these workload scales are such that conjoint analysis provides sensitivity to individual differences, and weighting provides better general prediction of experienced workload (Nygren, 1991). The simple SA(c) formula is sufficient for the intended general predictions of SA. In most practical field work, there is merit in simplicity and ease of implementation. Additional complexity and refinement should only be introduced if improvements are needed and shown to be beneficial.

SART couples ratings of knowledge with measurement of task demands and attentional resources, and in so doing includes factors associated

279

with workload (Selcon et al., 1991). Endsley (1993) argues that evidence for the dissociation of workload and SA calls for independent measurement of the two variables, and for attention to intervening factors that produce different SA/workload combinations. SA(c) provides a single estimate of SA influenced by factors affecting workload. But the workload factors are estimated by the Demand and Supply ratings, independently of the Understanding rating. Analysis of different SA and workload combinations can be made by examining the component levels of SA(c) and SART.

References

Crabtree M.S., Marcelo R.A., McCoy A.L. and Vidulich, M.A. (1993), " An examination of a subjective awareness measure during training on a tactical operations simulator", in Jenson, R. & Neumeister, D. (eds), *Proceedings of the 7th International Symposium on Aviation Psychology,* OSU, Columbus.

Endsley, M.R. (1993), "Situation awareness and workload: Flip sides of the same coin", in Jenson, R. & Neumeister, D. (eds), *Proceedings of the 7th International Symposium on Aviation Psychology,* OSU, Columbus.

Nygren, T.E. (1991), ' Psychometric properties of subjective workload measurement techniques: Implications for their use in the assessment of perceived mental workload', *Human Factors,* **33** (1), 17-33.

Selcon, S.J., Taylor, R.M. and Shadrake, R. (1992), 'Multi-modal cockpit warnings: Picture, words, or both?, in *Proceedings of the HFS 36th Annual Meeting,* **1**, 57-61, HFS, Santa Monica.

Selcon, S.J., Taylor, R.M. and Koritsas, E. (1991), 'Workload or situational awareness?: TLX vs SART for aerospace systems design evaluation', in *Proceedings of the HFS 35th Annual Meeting,* **1**, 62-66, HFS, Santa Monica.

Taylor R.M. (1990), 'Situation awareness rating technique (SART): The development of a tool for aircrew systems design', in *AGARD CP 478, Situation Awareness in Aerospace Operations,* AGARD, Neuilly sur Seine.

Taylor R.M. and Selcon, S.J. (1990) 'Evaluation of the situation awareness rating technique as a tool for aircrew systems design', in *AGARD CP 478, Situation Awareness in Aerospace Operations,* AGARD, Neuilly sur Seine.

Taylor, R.M. and Selcon, S.J. (1990), 'Cognitive quality and situational awareness with advanced aircraft attitude displays', in *Proceedings of the HFS 34th Annual Meeting,* **1**, 26-30, HFS, Santa Monica.

Vidulich, M.A., Crabtree M.S. and McCoy, A.L.(1993). " Developing subjective and objective metrics of pilot situation awareness", in Jenson, R. and Neumeister, D. (eds), *Proceedings of the 7th International Symposium on Aviation Psychology,* OSU, Columbus.

42 The role of Crew Resource Management (CRM) in achieving team situation awareness in aviation settings

Michelle M. Robertson, Institute of Safety and Systems Management, University of Southern California, Los Angeles, California
Mica R. Endsley, Department of Industrial Engineering, Texas Tech University, Lubbock, Texas

Crew resource management (CRM) training has gained widespread popularity in a variety of aviation settings since its inception in the late 70's and early 80's. Based on investigations of aircraft accidents, performance problems, simulator studies and comments from aircrew on inadequacies in current training processes, CRM set out to provide flight crews with training in the areas of leadership and command, interpersonal skills and communications, problem solving, and the management of limited resources, particularly under stress (Lauber 1987). More recently CRM programs have been extended from the cockpit to cabin crews and maintenance operations (Taylor, Robertson et al. 1993). Although CRM currently suffers from variability in its implementation and level of impact both between and within organizations (Helmreich and Wilhelm 1991; Helmreich 1993), CRM has been shown to have considerable impact on crew attitudes (Helmreich and Wilhelm 1991). These changes have been shown to be both lasting and related to aircrew behaviors (Butler 1991; Clothier 1991) and objective measures of effectiveness in aircraft maintenance technical operations (Taylor, Robertson et al. 1992).

Three primary crew attitudes have been found through factor analysis of the Cockpit Management Attitudes Questionnaire (CMAQ) to be affected by aircrew coordination training: *Coordination and communication, recognition of stressor effects, and command responsibility* (Helmreich, Foushee et al. 1987) In addition, Taylor et al. (1993) have found *avoiding conflict (*a reflected index of *willingness to voice disagreement)* to be statistically strong in measurements of maintenance crew attitudes. All four of these attitudes have been found to be positively related to self-reported behaviors and performance in the maintenance area.

In examining how CRM training's effect on attitudes may influence crew performance, a model of dynamic crew decision making is adopted in which situation awareness (SA), decision making, and performance are each viewed as separate stages that are impacted by different features of the system and the individual (Endsley 1988, in press). Although CRMs focus on decision making is often discussed, it is the objective of this paper to examine the ways in which it can impact the SA portion of the decision process.

Situation Awareness

The importance of SA in effective decision making has received considerable attention in recent years. A great deal of crew members' activities revolve around forming a correct perception and interpretation of the current situation. In complex, dynamic, and uncertain environments, such as the cockpit, this can be quite challenging. Formally defined, SA is "the perception of the elements in the environment within a volume of time and space, the comprehension of their meaning and the projection of their status in the near future" (Endsley 1988). As such, it embodies not only perceiving the relevant information (level 1), but also comprehending the meaning of that information, particularly when integrated and viewed in light of operational goals (level 2), and, based on this, projecting likely future events and scenarios (level 3). This high level understanding and prediction allows decision makers the opportunity to act in a timely manner to achieve goals. Without an accurate understanding of the situation, even the best decision makers may act inappropriately. Thus, SA can be viewed as critical in the decision process.

Given the same system design, different individuals have been shown to vary in their ability to achieve SA (Endsley and Bolstad 1993). This may be due to inherent abilities, experience and/or training, all of which combine to form various cognitive mechanisms used for acquiring and maintaining SA. Hypothesized mechanisms include: limited attention and working memory capacity; long-term memory stores, particularly mental models, which can circumvent these limits by facilitating comprehension and projection on the basis of even limited information; and goals and expectations which will influence how attention is directed, how information is perceived and how it is interpreted (Endsley, in press).

While SA has been discussed mainly at the level of the individual, it is also relevant for the aircrew as a team. *Team SA* has been defined as "the degree to which every team member possesses the SA required for his or her responsibilities" (Endsley 1993). For instance, if one crew member has a certain piece of information, but another who needs it does not, the SA of the team has suffered and their performance may suffer as well unless the discrepancy is corrected. In this light, a major portion of inter-crew coordination can be seen as the transfer of information from one crew member to another, as required for developing SA across the team. This coordination involves more than the passage of data to include a sharing of higher level SA.

This process is greatly enhanced by the creation of a *shared mental model* which provides a common frame of reference for team member actions and allows team members to predict each other's behaviors (Orasanu 1990; Cannon-Bowers, Salas et al. 1993). A shared mental model may provide more efficient communications by providing a common means of interpreting and predicting actions based on limited information, and therefore may form a crucial foundation for effective teamwork. For instance, Mosier and Chidester (1991) found that better performing teams communicated less than poorer performing teams.

Impact of CRM on SA

While there is some recognition of the importance of SA on team performance within the CRM literature (Prince and Salas 1993), a greater focus on this issue within the evolving concept of CRM is needed. We propose that CRM can impact crew SA by directly affecting individual SA, or indirectly through the development of shared mental models and by providing efficient distribution of attention across the crew. This will be discussed in light of the behaviors measured by the Line/LOS Checklist (LLC), as shown in Figure 1, which have been shown to be positively impacted by CRM (Butler 1991; Clothier 1991).

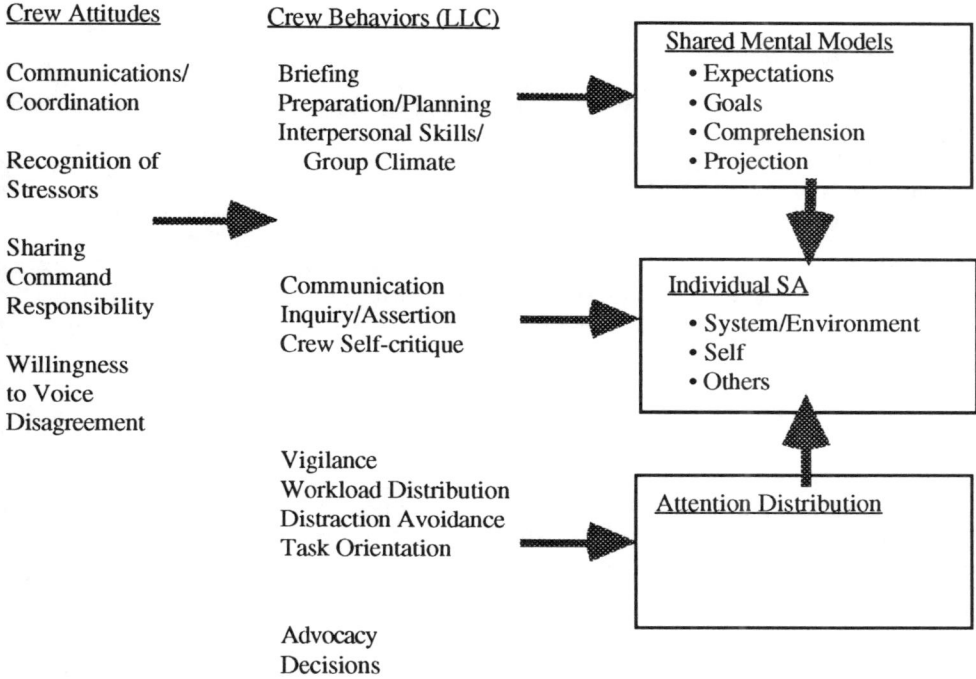

Crew Attitudes   Crew Behaviors (LLC)

Communications/   Briefing
Coordination   Preparation/Planning
                Interpersonal Skills/
Recognition of   Group Climate
Stressors

Sharing
Command   Communication
Responsibility   Inquiry/Assertion
                Crew Self-critique
Willingness
to Voice
Disagreement

                Vigilance
                Workload Distribution
                Distraction Avoidance
                Task Orientation

                Advocacy
                Decisions

Shared Mental Models
- Expectations
- Goals
- Comprehension
- Projection

Individual SA
- System/Environment
- Self
- Others

Attention Distribution

Figure 1 CRM related behaviors potentially impacting SA

Individual SA

Probably the factor most obviously affecting SA, improved *communication* between crew members facilitates effective sharing of needed information. In particular, *inquiry and assertion* help to insure this communication, which can include data on the system or environment, but also may include individual crew members' higher level assessments of the situation. In addition, an understanding of the state of the human elements in the system also forms a part of SA. Good *self-critique* provides an up-to-date assessment of one's own and other team member's abilities and performance, which may be impacted by factors such as fatigue or stress. This knowledge allows team members to recognize the need for providing more information and taking over functions in critical situations, an important part of effective team performance.

Shared mental models

The crew *briefing* establishes the initial basis for a shared mental model between crew members, providing shared goals and expectations. This can increase the likelihood that two crew members will form the same higher levels of SA from low level information, improving the effectiveness of communications. Prior *preparation and planning* similarly can help establish a shared mental model.

283

Effective crews tend to "think ahead" of the aircraft, allowing them to be ready for a wide variety of events. This is closely linked to Level 3 SA — projection of the future. An understanding of *interpersonal relationships and group climate* can facilitate the development of a good model of other crew members. This allows individuals to predict how others will act, forming the basis for Level 3 SA and efficiently functioning teams.

Attention Distribution

The effective management of human resources becomes extremely critical in high task load situations. A major factor in effectively managing these resources is ensuring that all aspects of the situation are being attended to — avoiding attentional narrowing and neglect of important information and tasks. *Task orientation* and *distribution of tasks under workload* directly impact how individuals within the team are directing their attention, and thus their SA. In addition, improvements in *vigilance* and the *avoidance of distractions* can be seen to directly impact team members' attention distribution. (*Advocacy* and *Decision* skills, while extremely valuable outcomes of CRM are viewed as primarily impacting decision making rather than SA.)

CRM and SA: Training implications

Given that CRM can effect crew SA in a number of ways and that SA is critical for effective performance, it is proposed that the potential impact of CRM on SA be strengthened through the specific incorporation of SA-oriented training in the CRM process. At the team level this could include (1) training that highlights the role of SA in the operational environment and the impact of CRM attitudes on SA, and (2) training of specific skills and behaviors that can bring about improved SA in the crew coordination/communication process.

Numerous portions of the CRM process have been shown to be linked to behaviors that impact on SA. This natural linkage should be brought out in existing CRM programs by highlighting the role of SA in the team decision and performance process and influence of CRM behaviors on SA. For instance, training programs for improving the quality of the shared mental model can be developed. In addition to shared experience, it may be possible to train crews to specifically work to improve their models of each other through communications, briefing, and prior preparation and planning during CRM. Specific behaviors must be identified, demonstrated and taught in CRM that will address this need.

Recommendations to incorporate training of specific behaviors, in addition to attitudes, have recently been made by several researchers (Helmreich 1993; Prince, Chidester et al. 1993). Towards this goal, team behaviors directed at enhancing SA should be identified. Prince and Salas (1993) have identified seven skills that they believe are important to the SA component of CRM: (1) identify problems/ potential problems, (2) recognize the need for action, (3) attempt to determine cause of discrepant information before proceeding, (4) provide information in advance, (5) note deviations, (6) demonstrate ongoing awareness of mission status, and (7) demonstrate awareness of task performance of self. This list provides a good starting point for the development of an SA-oriented CRM program.

Determining the most effective method for training these skills in CRM needs to be determined. Some essential components should include *providing skill information, demonstration of the skill, practice* and *feedback*. In addition, Dyer (1984) suggests training strategies for CRM include *behavioral modeling*. The development of *meta-cognitive skills* for self-monitoring and awareness of one's performance would seem to greatly support active practice and feedback and be essential for improving SA in the CRM process. In addition, the *ongoing maintenance* of these skills needs to be addressed, as a single program of training cannot assure that participants achieve and maintain the desired competence.

Systematic evaluation of CRM training

Assessing the impact of CRM programs on actual team performance is critical to their development and use in organizations. In the context of issues addressed in this paper, the program's impact on *team SA* can also be directly assessed through objective methods for assessing individual SA such as SAGAT (Endsley 1989).

To systematically evaluate the impact of any CRM training program on team performance, a multi-variate, longitudinal evaluation process is essential. A *formative evaluation process* provides a comprehensive framework for systematically evaluating the training program at several hierarchical levels, providing trainers and management with pertinent information on the effectiveness and efficiency of a training program during its development. Kirkpatrick (1976) proposed that evaluation include: *reactions to training, extent of learning, behavioral change and demonstration of transferred skills, and organizational performance and effectiveness.* In addition, Cannon-Bowers et al. (1989) identified *content validity and re-training assessment* as important components. Determining if objectives are met, the appropriateness of the training content and methods, maximization of transfer of training, and suggestions for program improvement all need to be assessed during the formative evaluation process.

Demonstration of systematic evaluation models are still sparse in the CRM training area, however Taylor et al. (1993) have applied all levels of this evaluation framework, and consequently were able to demonstrate the effectiveness of a CRM training program for aircraft maintenance operations. Reactions to the CRM training program have been shown to be favorable and positive with significant positive attitudinal changes measured immediately after, 2, 6, and 12 months following training. Maintenance performance measures in the areas of safety, dependability and efficiency were shown to be significantly correlated with the attitudinal data. This systematic, longitudinal data collection allows for a robust measure of the impact of CRM training. By applying this evaluation framework to efforts to incorporate SA into CRM programs, an adequate measure of the development, transfer and application of SA skills in the operational environment can be assessed over time.

References

Butler, R. E. (1991), 'Lessons from cross-fleet/cross airline observations: Evaluating the impact of CRM/LOS training', *Proceedings of the Sixth International Symposium on Aviation Psychology*, Department of Aviation, The Ohio State University, Columbus, OH.

Cannon-Bowers, J. A., Prince, C. W., Salas, E., et al. (1989), 'Determining aircrew coordination training effectiveness', *Proceedings of the Eleventh Interservice/Industry Training Systems Conference (I/ITSC)*, National Security Industrial Association, Fort Worth, TX.

Cannon-Bowers, J. A., Salas, E. and Converse, S. (1993), 'Shared mental models in expert team decision making', in N. J. Castellan (eds), *Current issues in individual and group decision making*, Lawrence Earlbaum, Hillsdale, NJ.

Clothier, C. (1991), 'Behavioral interactions in various aircraft types: Results of systematic observation of line operations and simulations', *Proceedings of the*

Sixth International Conference on Aviation Psychology, Department of Aviation, The Ohio State University, Columbus, OH.

Dyer, J. (1984), 'Team research and team training: A state of the art review', in F. Muckler (ed), *Human Factors Review*, Human Factors Society, Santa Monica, CA.

Endsley, M. R. (1988), 'Design and evaluation for situation awareness enhancement', *Proceedings of the Human Factors Society 32nd Annual Meeting*, Human Factors Society, Santa Monica, CA.

Endsley, M. R. (1989), 'A methodology for the objective measurement of situation awareness', in (eds), *Situational Awareness in Aerospace Operations (AGARD-CP-478)*, NATO - AGARD, Neuilly Sur Seine, France.

Endsley, M. R. (1993,February), 'Situation awareness in dynamic human decision making: Theory', *Presented at the First International Conference on Situational Awareness in Complex Systems*, Orlando, FL.

Endsley, M. R. (in press),'Towards a theory of situation awareness', *Human Factors*.

Endsley, M. R. and Bolstad, C. A. (1993), 'Human capabilities and limitations in situation awareness', in (eds), *Combat Automation for Airborne Weapon Systems: Man/Machine Interface Trends and Technologies (AGARD-CP-520)*, NATO - AGARD, Neuilly Sur Seine, France.

Helmreich, R. L. (1993), 'Whither CRM? Future directions in crew resource management training in the cockpit and elsewhere', *Proceedings of the Seventh International Symposium on Aviation Psychology*, Department of Aviation, The Ohio State University, Columbus, OH.

Helmreich, R. L., Foushee, H. C., Benson, R. and Russini, W. (1987), 'Cockpit management attitudes: Exploring the attitude behavior linkage', *Aviation, Space and Environmental Medicine*, **57**1198-1200.

Helmreich, R. L. and Wilhelm, J. A.(1991),'Outcomes of crew resource management training', *International Journal of Aviation Psychology*, **1**(4): 287-300.

Kirkpatrick, D. L. (1976), 'Evaluation of training', in R. L. Craig (ed.), *Training and development handbook: A guide to human resource development*, McGraw Hill, New York.

Lauber, J. K. (1987), 'Cockpit resource management: Background and overview', in H. W. Orlady and H. C. Foushee (eds), *Cockpit resource management training: Proceedings of the NASA/MAC workshop (NASA Conference Publication #2455)*, NASA-Ames Research Center, Moffett Field, CA.

Mosier, K. L. and Chidester, T. R. (1991), 'Situation assessment and situation awareness in a team setting', in Y. Queinnec and F. Daniellou (eds), *Designing for Everyone*, Taylor and Francis, London.

Orasanu, J. (1990,July), 'Shared mental models and crew decision making', *12th Annual Conference of the Cognitive Science Society*, Cambridge, MA.

Prince, C., Chidester, T. R., Bowers, C. and Cannon-Bowers, J. (1993), *Research in training for teamwork in the military aircrew,* in E. Wiener, B. Kannki and R. Helmreich (eds), *Cockpit Resource Management*, Academic Press, San Diego, CA.

Prince, C. and Salas, E. (1993), 'Training and research for teamwork in the military aircrew', in E. Wiener, B. Kannki and R. Helmreich (eds), *Cockpit Resource Management*, Academic Press, San Diego, CA.

Taylor, J. C., Robertson, M. M., Peck, R. and Stelly, J. W. (1993), 'Validating the impact of maintenance CRM training', *Proceedings of the Seventh International Symposium on Aviation Psychology*, Department of Aviation, The Ohio State University, Columbus, OH.

43 A taxonomy of situation awareness errors

Mica R. Endsley, Department of Industrial Engineering, Texas Tech University, Lubbock, Texas

In the dynamic flight environment effective aircrew decision making is highly dependent on situation awareness (SA) — a constantly evolving picture of the state of the environment. SA is formally defined as a person's "perception of the elements in the environment within a volume of time and space, the comprehension of their meaning, and the projection of their status in the near future" (Endsley, 1988). It encompasses not only an awareness of specific key elements in the situation (Level 1 SA), but also a gestalt comprehension and integration of that information in light of operational goals (Level 2 SA), along with an ability to project future states of the system (Level 3 SA). These higher levels of SA (2 & 3) are particularly critical to effective functioning in complex environments, such as the cockpit.

The failures of human decision making are frequently cited in investigations of error in a wide variety of systems. In aviation mishaps, failures in decision making are attributed as a causal in factor in approximately 51.6% of all fatal accidents and 35.1% of non-fatal accidents, of the 80-85% of accidents which are attributed to human error (Jensen, 1982). It is the central thesis of this paper that while some of these incidents may represent failures in actual decision making (action selection), a high percentage of these errors are actually errors in situation awareness. That is, the aircrew makes the correct decision for their perception of the situation, but their perception is in error. This represents a fundamentally different category of problem than a decision error — in which the correct situation is comprehended, but a poor decision is made as to the course of action to take — and indicates different types of remediation attempts.

Shortcomings in aircrew SA have been directly linked to decision making and performance errors. Hartel, Smith and Prince (1991) found SA to be the leading causal factor in a review of 175 military aviation mishaps. SA break-downs can occur due to either *incomplete SA* — knowledge of only some of the elements — or *inaccurate SA* — erroneous knowledge concerning the value of some elements or the integration and comprehension of those elements.

Based on a review of literature on human information processing and cognition, a taxonomy for classifying and describing errors in SA has been developed (Endsley, 1993; in press). The taxonomy, presented in Table 1, incorporates factors affecting SA at each of its three levels. In addition, several general factors impacting SA across its three levels are presented. (The taxonomy will be described in more detail subsequently.) In order to further develop the taxonomy and to use it to better understand the factors leading to SA errors, a review of major aviation accidents was conducted.

Table 1. SA error taxonomy

SA Error Causal Factor Category	Freq
I. Level 1 SA - Failure to Correctly Perceive Situation	23
A. Data not available	3
B. Data difficult to detect/perceive	5
C. Failure to scan or observe data	
1. Omission	3
2. Attentional narrowing/distraction	3
3. High taskload	4
D. Misperception of data	4
E. Memory failure	1
2. Level 2 SA - Failure to Comprehend Situation	7
A. Lack of/poor mental model	1
B. Use of incorrect mental model	2
C. Over-reliance on default values in model	1
D. Memory failure	0
E. Other	3
3. Level 3 SA - Failure to Project Situation into the Future	2
A. Lack of/poor mental model	1
B. Other	1
Maintaining Multiple Goals	0
Habitual Schema	0

Accident Investigation

Major aircarrier accidents in the United States were reviewed through an analysis of National Transportation Safety Board (NTSB) accident investigation reports from the years 1989 - 1992. In total this included reports on 24 accidents that were available in Regional Government Document Repositories. Of these, 9 (38%) involved mechanical failures, 8 (33%) involved inclement weather, and 17 (71%) involved some type of inflight human error on the part of the aircrew (13), air traffic controllers (2) or both (2). (Human errors associated with maintenance operations possibly leading to mechanical failure were not addressed in this study.)

The 17 accidents which involved some type of human error were further investigated for causal factors. (While this number is too small for reliable statistics to be drawn, the accidents do provide good descriptive illustrations of the types of factors which can contribute to human error, and more specifically situation awareness errors.)   Five categories of causal factor for human error were established: decision making, SA, physiological, procedural, and psycho-motor.

By far the most prevalent causal factor was some error in the aircrew's or controller's *situation awareness* (15 cases).   Many accidents also involved other causal factors that contributed to the SA error.   Four of the accidents involved some sort of *physiological* degradation, usually fatigue (in one case drugs), which was cited by the NTSB as a contributory factor, and which could be directly related to a failure to achieve SA.   Six cases involved a *procedural* error, in which a violation of existing procedures occurred, setting up the crew for an SA error.   This primarily involved omitting a task (frequently a step on the checklist) or misexecuting a task due to inexperience.   Only two accidents presented clear evidence of a problem with *psychomotor* skills — basic control of the aircraft — as being a primary causal factor.   In four of the accidents there was reasonable evidence that the aircrew had accurate information on the situation, but made a poor *decision* none-the-less.   This included proceeding into inclement weather, neglecting to go IFR when conditions warranted, declining de-icing, and continuing on a missed approach beyond the point dictated by procedures.   In several cases, the poor SA that led to eventual accident was induced by these earlier decisions (e.g. continuing into weather and staying VFR in poor conditions).

The 15 accidents in which an SA error occurred will be described in more detail.   In many of these accidents multiple SA errors were present, involving SA errors on the part of multiple crew members and/or controllers or multiple SA errors (at different points in time) for the same crew member, creating a total of 32 SA errors.   Twenty-three (72%) involved Level 1 SA errors, failure to correctly perceived some pieces of information in the situation.   Seven (22%) involved a Level 2 error in which the data was perceived but not integrated or comprehended correctly.   Only two (6%) involved a Level 3 error in which there was primarily a failure to properly project the near future based on the aircrew's understanding of the situation.   The low number may be partially due to the difficulty of making such an assessment based on the limited data available following accidents.   (An error was always rated at its lowest base level — e.g. a lack of awareness of altitude was rated as a Level 1 error, even though it led to a lack of understanding of flight safety (Level 2) and prediction of time until crashing (Level 3).)

Table 1 includes a summary of the number of times each causal factor was identified in the 32 SA errors.   As with most accidents, multiple causal factors were often present.   For example,   there were frequently several opportunities for a person to acquire a piece of information and there may have been different reasons for the error in each case. In some cases, although they recovered from the first SA error, a second SA error occurred leading to the accident. An overview of the accidents and causal factors is presented in Table 2. A detailed discussion of the SA error taxonomy, illustrated by examples from these errors follows.

SA Error Taxonomy

Level 1 - Failure to correctly perceive the situation

At the most basic level, important information may not be correctly perceived. The data may *not be available* to the person, due to a failure of the system design to present it or a failure in the communications process.   In two of the accidents reviewed, automation systems failed to present a signal to the crew that a door was not properly latched or the flaps were not properly set.   In another case, a controller kept an aircraft in a repeated series of holding patterns until the aircraft

289

Table 2. Summary of causal factors in each accident involving SA errors

Accident	Description	Position	Causal Factors	Other Factors
NTSB/AAR-92/05	Spatial disorientation	Captain	Level 1 - misperception	Physiological
NTSB/AAR-92/01	Crash on approach	Captain	Level 2 - other (significance)	Weather, Decision
NTSB/AAR-91/09	Crash on take-off (icing)	Captain	Level 2 - no/poor model	Weather, Procedure, Physiological
NTSB/AAR-91/08	Landed aircraft on occupied runway	Controller	Level 1 - difficult to detect, memory failure/taskload, distraction	
NTSB/AAR-91/05	Runway collision (fog)	Crew	Level 1 - difficult to detect	Weather, Procedure
		Controller	Level 1 - difficult to detect	
			Level 2 - over-reliance on defaults	
NTSB/AAR-91/04	Ran out of fuel	Crew	Level 2 - other (significance)	Weather
			Level 3 - failure to project (time)	
		Controller	Level 1 - no data	
NTSB/AAR-91/03	Landed aircraft on occupied runway	Controller 1	Level 2 - other (integration)	
		Controller 2	Level 1 - failure/distraction	
NTSB/AAR-91/01	Loss of control - landing	First Officer	Level 1 - failure/omission	Psycho-motor
NTSB/AAR-90/05	Crash into mountain	Crew	Level 1 - difficult to detect	Weather, Decision
			Level 2 - wrong model	Physiological
NTSB/AAR-90/04	Struck power lines	Crew	Level 1 - misperception	Weather, Procedure
NTSB/AAR-90/03	Crash on take-off (mistrimmed rudder)	First Officer	Level 1 - failure/taskload	Procedure
		Captain	Level 2 - wrong model	
NTSB/AAR-90/02	Loss of control (cargo door open)	First Officer	Level 1 - difficult to detect	Mechanical
		Captain	Level 1 - no data (auto. failure)	Psycho-motor
		Captain	Level 3 - no/poor model	
NTSB/AAR-89/04	Crash on take-off (mis-set flaps & slats)	First Officer	Level 1 - failure/taskload	Procedure
		Captain	Level 1 - no data (auto. failure)	
NTSB/AAR-89/01 (sum)	Crash on approach	First Officer	Level 1 - failure/taskload, misperception	Decision
		Captain	Level 1 - failure/attn. narrowing	
			Level 1 - failure/misperception	
NTSB/AAR-89/01	Crash into mountain	First Officer	Level 1 - failure/omission	Procedure
		Captain	Level 1 - failure/omission	Physiological

290

reached an emergency fuel condition and crashed on approach. The controller was never notified by the crew of any fuel problem until it was critical. In other cases, the data is available, but is *difficult to detect or perceive.* In the cases reviewed, this included poor visual information in darkness and poor weather (fog, heavy rain), poor visual conditions created by glare from runway lighting, and design features that were difficult to discriminate (door latching mechanisms).

Many times, the information is directly available, but is *not observed or included in the scan pattern.* This may occur for any of several reasons. A simple *omission* occurred in two cases where a step on the checklist was skipped or misexecuted. In another case, an inexperienced crew member did not set up a good instrument scan pattern. In other cases, individuals fell victim to *attentional narrowing or other distractions* that prevented them from attending to important information. In several cases, controllers were distracted by communications with other aircraft, causing them to neglect attending to the aircraft involved in the accident. In another case, the aircraft captain became fixated on the flight direction indicator, ignoring other basic information which would have corrected his misunderstanding of the situation. *High taskload* (even momentary) can also prevent information from being attended to. High taskload was attributed as causal in several cases in which checklists were hurried through, or individuals were busy with many things, leading to poor attendance to any one task.

In other cases, information is attended to, but is *misperceived.* In many cases, this was directly due to the person's expectations. They expected to see a certain thing and misperceived environmental cues in line with those expectations. For example, one crew arriving on an unfamiliar approach in poor visibility mistook the lights from a car dealership as the runway lights. The misperception of the lights led them to believe they were closer to the runway than they were and consequently flew much too low and struck power lines. In other cases, information was simply misread or misinterpreted, or misperceived due to spatial disorientation (conflicting physiological cues).

Finally, in some cases it is apparent that a person initially perceives some piece of information but then forgets about it. Situation awareness often involves keeping information about a large number of factors in memory. *Memory failure* was a major reason for one accident in which a controller landed an incoming aircraft on a runway which she had previously assigned another aircraft to. (Contributing factors included poor visibility, high taskload and distractions.)

Level 2 SA - Failure to comprehend the situation

In other cases, information is correctly perceived, but its significance or meaning is not comprehended. This may be due to the *lack of a good mental model.* In one case, a captain refused de-icing because he had no information on the particular susceptibility of that model of aircraft to icing problems in certain conditions. In other cases, the *wrong model* may be used to interpret information. The captain of one aircraft which attempted to take-off without the flaps set correctly was aware of several cues that should have indicated there was a problem (misaligned rudder pedals, steering drift to the left), but did not comprehend their meaning because he relied on his model of how another, more familiar aircraft functioned. In another case, a pilot using his model of the geography of one part of an island interpreted visual cues into that model. This led him to run into a mountain that he did not expect to be there and could not see.

It is also possible for a person to have a good model of how a system functions, but *over-rely on default values* in the model. These defaults can be thought of as general expectations about how parts of the system function that may be used in the absence of real-time data. For example in one case, a

291

controller assumed a particular aircraft had taken-off based on general knowledge about how long after a clearance aircraft usually take before departing. (Direct visual information was obscured.) Consequently the controller did not notify the aircraft about the presence of another aircraft which was lost in the fog nearby.

In other cases, the significance of perceived information relative to operational goals is simply not comprehended or several pieces of information are not properly integrated. This may be due to *working memory limitations* or *other* unknown cognitive lapses. In one case, the controller brought in several aircraft with close spacing and overtaking velocities. This led to landing an aircraft on an occupied runway. In another case, a pilot apparently under-appreciated the severity of a weather situation, even though other aircraft were executing go-arounds. The crew of the aircraft that ran out of fuel apparently did not appreciate the severity of the situation until it was too late either. No known reason really explains these lapses by trained individuals.

Level 3 SA - Failure to project situation into the future

Finally, in some cases individuals may be fully aware of what is going on, but have a *poor model* for projecting what that means for the future. In one case in which a cargo door opened in flight, the crew still could have landed safely, however they did not have a good model of how the open cargo door would affect the handing characteristics of the aircraft. The captain executed a sharp turn which caused the door to swing outward more, leading to an unrecoverable, usual attitude the captain was not attending to as he looked outside for runway lights. With a better model he would have been able to project the effect of this maneuver and been ready to monitor his attitude more carefully. In *other* cases, the reason for not correctly projecting the situation is less apparent. The crew of the aircraft that ran out of fuel probably did not correctly project (soon enough) how much time they needed to land safely, even though they should have been able to. Mental projection is a very demanding task at which people are generally poor.

In addition to theses main categories, two general categories of causal factors are included in the taxonomy. First some people have been found to be poor at *maintaining multiple goals* in memory, which could impact SA across all three levels. Secondly, there is evidence that people can fall into a trap of executing *habitual schema*, doing tasks automatically, which render them less receptive to important environmental cues. None of the cases reviewed provided any clear evidence for either of these categories.

References

Endsley, M. R. (1988), 'Design and evaluation for situation awareness enhancement', *Proceedings of the Human Factors Society 32nd Annual Meeting*, Human Factors Society, Santa Monica, CA.

Endsley, M. R. (1993,February), 'Situation awareness in dynamic human decision making: Theory', *Presented at the First International Conference on Situational Awareness in Complex Systems*, Orlando, FL.

Endsley, M. R. (in press), 'Towards a theory of situation awareness', *Human Factors*.

Hartel, C. E., Smith, K. and Prince, C. (1991,April), 'Defining aircrew coordination: Searching mishaps for meaning', *Presented at the Sixth International Symposium on Aviation Psychology*, Columbus, OH.

Jensen, R. S. (1982), 'Pilot judgment training and evaluation', *Human Factors*, 24(1): 61-73.

44 Distributed situation awareness: a concept to cope with the challenges of tomorrow

Denis Javaux*, Work Psychology Department, University of Liege, Belgium
Sylvie Figarol, CENA, Athis-Mons, France

*supported by the National Belgian Policy, PAI

An intuitive understanding of Situation Awareness (SA) could refer to a mental state or process dealing with the assessment of multiple dynamic elements varying while performing a task. Whatever the viewpoint, SA clearly seems to resort to cognition, and it has been recognized as "an essential prerequisite for the safe operation of any complex dynamic system" (Sarter & Woods, 91).

Since several important changes on the flight deck are expected in the near future (e.g. datalink) and might impact dramatically the way SA is built and circulated, we believe that the concept of SA should be revisited or extended. As an illustration, it will be presented, through the description of a specific flight phase, and some issues of SA assessment will be discussed in the context of data-link design.

Components of SA

Classical acceptations consider SA as an awareness of a problem-space under control. We believe that such a definition is simply inoperative for understanding the flying task, related activities, or cognitive resources implied in its achievement.

293

Whatever the agent in charge of the plane, it has to cope with changing constraints in the environment (weather, traffic,...), changing resources (fuel consumption, engine failures,...), a changing task (while still being partially undefined on take-off, it is progressively constrained and specified as the flight goes on and more and more information about unpredictable aspects is obtained),...

Problem-space or system awareness is hence not sufficient for an agent to achieve the task. Task and resources awareness are necessary, especially in situations where the two components are changing rapidly. We thus argue that *three essential components of SA* must be considered for understanding flying task related operations : **Task, Resources and System awareness.** Moreover, each of the components of SA must be considered *on a temporal dimension*, as stated by Sarter & Woods (1991).

The nature of the cognitive agent SA belongs to

The classical position states that the pilots are flying the airplane (Billings, 91), and this is coherent with the classical view of SA, considered as a mental state or process of the pilots. We think this position is too narrow : the pilots are not alone in charge of the flying task.

An inclusion criteria can be defined : the crew (flying the airplane) is composed of all the agents whose activity takes part to the realization of the (flying) task. According to this criteria, the crew involved in the control loop over the airplane is composed of the two (or three) cockpit crew members, to whom we have to add a) the air traffic controllers, since they get an awareness of the relation between the airplane and its remote environment : weather, traffic, airport configuration... and b) the high level computer agents (FMS, AP, A/THR, ECAM/EICAS) since they get awareness of the plane at various levels and sometimes act on them (i.e. are involved in the control loop).

The airplane is thus flown by a distributed cooperative system acting as a whole : the **macro pilot** or **extended crew.** Moreover, since SA is a cognitive property of the agent in charge of the task, we must consider a higher level of SA : **Distributed Situation Awareness** (DSA) or SA of the macro-pilot. This extended view of SA is supported by the most recent trends in Cognitive Sciences (see for example Bond & Gasser, 88; Minski, 86; Hutchins & Klausen, 91). The macro-pilot must be understood as a single and complex agent, whose SA must be built and maintained in order to achieve the flying task.

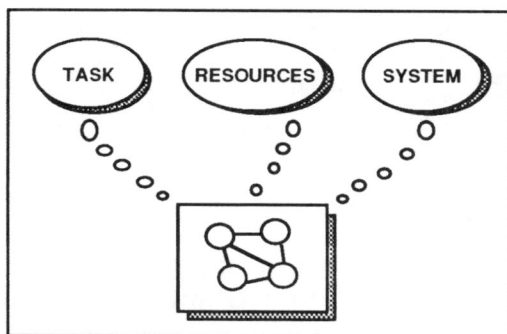

Fig. 1. Distributed crew and the three facets of SA

We believe that one can only understand how the local SA is built, maintained and circulated when relating it to the global SA and to the task of the macro-pilot. This point might be crucial since the internal structure of the macro-pilot is soon going to change. Designers will have to take care of these aspects when they will introduce new datalink capabilities.

Illustration of SA assessment in flying task

As an example, a partial description of the approach phase will lead to the illustration of the properties of SA as described previously and to the identification of some bottlenecks related to present SA building and circulation.

Description of the approach phase

The description of approach phase is based on **prescribed task** definition, and will focus on Task Awareness, which is a very important component, since it plays two main roles : temporally organise the actions, according to action objectives and events coming in, and constrain the envelope of possible trajectories by taking into account the available means of achieving the task.

We will try to assess when and how TA is created and adapted, which agents take part to this creation, and when and how TA is circulated between the different agents. The approach phase will be described on a temporal and chronological basis, and the temporal structure of TA will be emphasized.

What happens to TA during descent and approach ?

Before descent, PNF reads the ATIS message and transmits the data to both PF and FMS (via MCDU). This triggers a process of *TA creation* for approach :

1- FMS refines a flight plan, with a precise 4D trajectory, based on the input data (STAR, QFU, type of approach) : this flight plan from descent to landing, takes into account some resources constraints like fuel optimization, and integrates the branching point (MDA) between landing and Go-Around.

2- PF builds a plan including the envelope of possible trajectory : PF can forecast a certain descent profile towards TMA entry, but trajectory from TMA entry and axis interception remains uncertain. In the last landing segment, there is a branching with go-around trajectory. By reading the proposed flight plan on the MCDU, PF keeps *adapting* the TA, by adding constraints to the task. He may include resources and system awareness, and elaborate one or several strategies to perform the approach.
At this time, TA is global, completely created and adapted to context, according to the available data. TA is a sort of plan with a quite extended time horizon, and with a temporal branching structure : PF may have asked PNF to prepare a holding pattern, if he expects a possible waiting time because of high traffic, because of weather conditions (visibility), or because some specific system state. Or PF may have planned an alternative visual approach or runway to be requested to ATC when appropriate.
This *TA creation* is partly structured on board by the actions prescribed in the descent procedure, and is distributed over PF and FMS.

3- Once created and refined, *TA is circulated* from PF to PNF through the arrival briefing. This briefing normally takes place just before descent initialisation : its structure is described in the prescribed task and serves as a real efficient circulation support... In particular, the briefing includes :

- definition of an envelope of arrival trajectories : task is thus defined as constraints (altitude constraints, procedural constraints) and with branchings (linked to context conditions and to expected events) as described earlier.
- definition of constraints and objectives on system-related variables, according to system states or weather conditions,
- strategy of resources use (AP, A/THR, anti-icing , radionavigation aids, the type of display),
- task allocation between on-board agents : task sharing between PF and PNF, and FMS (choice for descent mode and navigation mode). PF is locally in charge of distributed SA, at this time.

4- Long after descent, TA for the approach is partially created on ground, when arrival controllers prepare arrival sequence, and roughly plan each aircraft trajectory from TMA entry to final approach.

5- Then this TA *is circulated* from the controllers to the pilots through instructions, in a partial way, and on a tactical basis.

6- Then during execution, the approach will be structured by standard procedures, and TA will be *adapted* to context according two ways :

- branching will be solved with expected events occurring : this branching, with the related definition of pivot, allow anticipation.
- unexpected events will trigger a particular adaptation of the task, sometimes loosing the constraints, or switching to another procedure, as described in Figarol & Javaux (1993).

What are the main bottlenecks in these processes of TA creation and circulation ?

From a temporal viewpoint, TA creation on board takes place long before TA creation on ground, and there is neither temporal synchronization, nor objective coordinations.

From a crew cooperation viewpoint, TA is circulated through vocal air-ground exchanges, on a tactical basis, which is not in favour of a strategic flight management.

Conclusion

This type of description stresses the dynamic SA updating and circulating process. It particularly points out the importance of SA building and the temporal dimension of SA, as it shows off the important Task Awareness component. This TA component is one condition of SA enhancement, since it allows anticipation and means-ends definition. It appears that in present flying situation, briefings and procedures are good tools for TA circulation.

Moreover, with unexpected events coming in, resources and system awareness are necessary to switch from procedural level to a higher level of constraints (TA adaptation). The three components of SA seem to be relevant.

Another result from this approach is that present aeronautical system generates "holes" in SA, on both air and ground sides : pilots are not informed of controllers'constraints and intents. This implies a complex temporal structure when creating their TA. On another side, controllers would probably enhance their SA with information about on board constraints and resources.

The diagnosis is that SA is presently available and distributed among the macropilot but it is circulated in a too partial way, and air ground integration could be designed to enhance this circulation.
This approach is promising and would require to be enriched by including ATC task assessment.

References

Billings C.E. (1991). *Human-Centered Aircraft Automation*. NASA Tech. memo N¡ 103885. Moffett Field, CA : NASA-Ames Research Center.

Bond, A.H., Gasser, L. (1988). (Eds). *Readings in Distributed Artificial Intelligence*, Morgan Kaufman.

Figarol, S., Javaux, D. (1993). *From Flying Task Description to Safety Assessment*. Le Transpondeur, dec 93.

Hutchins, E. and Klausen, T. (1991) *Distributed Cognition in an Airline Cockpit*, in Cockpit cognition.

Minsky, M. (1986). *The Society of mind*. N.Y. Simon and Schuster

Sarter N.B. and Woods D.D. (1991). *Situation awareness : A critical but ill-defined problem*. The International Journal of Aviation Psychology, **1** , 45-57.

45 Situational awareness in US Air Force F-15 pilots: no substitute for experience

Thomas R. Carretta, David C. Perry, Jr and Malcolm Ree
Armstrong Laboratory

Aircraft accidents are often blamed on a lack of situational awareness. The ability of a pilot to know location in space and time, and to keep track of other aspects of the dynamic environment of flight, are the common elements of definitions of situational awareness (SA). Despite the commonality, there is little agreement. Definitions range from "a sixth sense" (Hartman & Secrist, 1991) to "mental representations of various flight relevant dimensions" (Andre, Wickens, Moorman, & Boschelli, 1991). Even without a consensus most pilots consider SA sufficiently important to warrant research.

The impetus for this study was provided by the U S Air Force Chief of Staff. He has directed U S Air Force (USAF) research laboratories to investigate human attributes enabling a pilot to develop and maintain SA, especially in the high performance F-15 jet aircraft.

Both multiple empirical studies and meta-analyses have shown three predictors that are valid for almost all job performance criteria. These are general cognitive ability, *g* (Hunter & Hunter, 1984; McHenry, Hough, Toquam, Hanson, & Ashworth, 1990; Ree & Earles, 1992), psychomotor skill (Hunter & Hunter, 1984), and the personality construct of "conscientiousness" (Tett, Jackson, & Rothstein, 1991). The USAF currently collects measures of general cognitive ability and psychomotor ability in the Air Force Officer Qualifying Test (AFOQT) and the Basic Attributes Test (BAT; Carretta & Ree, in press).

This concurrent validation study (correlation design) was conducted from a personnel selection standpoint. The purpose was to determine which human attributes were predictive of SA.

Method

Subjects

The subjects were 171 USAF F-15 A/C pilots. These active duty pilots had between 1 and 20 years flying experience and between 88 and 2,007 F-15 flying hours and between 193 and 2805 total flying hours.

Measures

Predictors. The SA predictors were representative of general cognitive ability, psychomotor skills, and personality. At the first-order factor level there were cognitive measures of working memory, velocity estimation, near threshold processing, reasoning, and spatial. There were psychomotor measures of multilimb coordination, aiming, control precision, reaction time, and rate control. The Big 5 personality measures (Goldberg, 1990) were extroversion, emotional stability, agreeableness, conscientiousness, and openness. More detailed descriptions of the specific tests appear in Carretta, Perry, and Ree (in press). Cognitive, psychomotor, and personality measures were administered by a computer-based system (Carretta & Ree, in press).

Criteria. The criterion (Houck, Whitaker, & Kendall, 1991)was derived from multiple supervisory and peer SA ratings developed from task analyses with experienced F-15 pilots as subject matter experts. This resulted in 31 behavioral items representing personal traits and job tasks related to SA. Rating items represent general traits, tactical game plan, systems operation, communication, information interpretation, and tactical employment. Standardized definitions for each of the items were provided to every rater to establish consistency. Each of the 31 items was rated on a six point Likert scale from 1- "Acceptable" to 6- "Outstanding." All subjects were rated by multiple raters. This rating scale is provided in Carretta, Perry, and Ree (in press).

For peer ratings, pilots rated other pilots in their squadron with whom they had flown. Overall fighter ability and SA ability were scored on a six-point Likert scale ranging from 1- "Acceptable" to 6- "Outstanding." Pilots then rank-ordered peers from 1- "the best I've flown with " to N, the number of peers rated, indicating their judged standing on the trait of SA.

Control Variables. The consequences of job experience are important in understanding the relationship between ability and SA. This was accomplished by statistically holding flying experience constant (Schmidt, Hunter, &

300

Outerbridge, 1986).

Procedure

Subjects were tested on the computerized battery at their operational AFBs. Supervisory and peer ratings of SA were collected independently. The cognitive test scores which included both accuracy and time were formed into ratios to yield a measure of correct responses per unit time (Kyllonen, 1993).

Analyses

The issue of single versus multiple criteria was addressed by investigation of the unrotated SA ratings principal components. If the first unrotated principal component accounted for a large portion of the variance and the succeeding components a small proportion, a single criterion would be preferable.

The control variables were selected through regression analyses. Regressions of the criterion on F-15 hours, F-15 hours squared, F-15 hours cubed, total flying hours, total flying hours squared, and total flying hours cubed were computed. The squared and cubed terms were necessary to account for any non-linear relationships. Only those variables that contributed significantly to the regression were kept. Validity of the tests to predict SA was assessed with the effect of job experience controlled for statistically by partial correlation and by entering job experience control variables into the regression equations with test scores. Test scores were included in the regression equations if they showed a significant partial correlation with the criterion.

To understand the relationship among the predictors, principal components and factor analyses were conducted on all sets of predictors that showed a significant partial correlation (flying experience being held constant) with the criterion. The two predictor sets from which composites were formed included the cognitive tests and the psychomotor tests and were developed through partial correlation analyses.

The personality scales were not factor analyzed because they were already the consequence of factor theory.

Tests of linear regression models were used to assess the validity and incremental validity of the predictors. The first linear model (M1) was a full model that included experience control variables, a general cognitive ability composite (GCA), a psychomotor composite (PM), and a measure of Dependability (DEP). The personality construct of DEP was included because past research (Tett, et al., 1991) suggested it would be predictive. M1 also included interactions among cognitive, psychomotor, and personality composites.

The first reduced linear model (M2) was M1 with the 3-way interaction removed and similarly, M3 was M1 with all interactions removed. Reduced models M4, M5, and M6 contained the flying experience control variables and combinations of two of the three predictors found in model M3. Models M7, M8, and M9

301

contained the control variables and one of the predictor composites. M10 contained the control variables only. M7, M8, and M9 were tested against M10.

The statistical testing began by removing the interaction terms and then the individual GCA, PM, and DEP composites to determine which were statistically significant (p < .05) predictors of SA.

Results

Principal components analysis of SA ratings yielded only one eigenvalue greater than one. This factor accounted for 92.5% of the variance in the ratings demonstrating that a single criterion was preferable. Consequently, the first unrotated principal component of the ratings was the SA criterion. The two control variables found useful were F-15 flying hours and F-15 flying hours squared (R = .704)..

The matrix of partial correlations holding flying experience constant disclosed six tests as significant predictors of SA. These were measures of verbal working memory, spatial reasoning, divided attention, spatial working memory, aiming, and attention, reaction time, and rate control.

Principal components analysis of the four cognitive tests which had significant partial correlations showed that the first component accounted for 51% of the variance. The unit-weighted sum of these tests became the measure of GCA.

A principal components analysis of the two PM tests was not conducted as at least three variables are required. The unit-weighted composite of the two tests became the psychomotor composite.

As DEP was already based on factor analyses, the sum of the scale items became the DEP composite.

All linear regression models were significantly correlated with the criterion. Statistical tests showed that the interactions were not significant. This was determined by testing M2 against M1 and testing M3 against M2. To test the incremental validity of each type of predictor (GCA, PM, DEP), M4, M5, and M6 were tested against M3. No differences were found for M3 versus M4 or M3 versus M5. However, when GCA was removed, M3 versus M6, a significant difference was found. Further, M4 versus M7 and M5 versus M7 were tested and found not to differ. This showed that PM and DEP were not incremental to GCA for prediction of SA. Also, the comparisons for M7, M8, and M9 versus M10, revealed that only GCA provided incremental validity beyond flying experience.

Discussion

Flying experience in the F-15, which brings F-15 job knowledge, was the most predictive variable. This is consistent with Hunter's (1983) demonstration that ability influences job performance via the accumulation of job knowledge. The implication is that if pilots were allowed to acquire more flying hours, their job knowledge would be expected to increase as would their SA..

When job experience was held constant in the regressions, general cognitive ability was found to be predictive of the criterion. The psychomotor score and the personality trait of Dependability were not.

The results for general cognitive ability are in agreement with McHenry et al. (1990) who demonstrated that general cognitive ability was predictive of job performance in nine jobs. Ree and Carretta (in press) provide a broader discussion of the role of general cognitive ability in pilot selection. That the psychomotor composite was not predictive of SA when job experience was held constant was contrary to Carretta and Ree (in press). They demonstrated the incremental validity of psychomotor measures for predicting pilot training performance. Ree and Carretta (1992) have demonstrated that psychomotor tests measure general cognitive ability, along with general and specific psychomotor ability. It is likely that the constant training provided by frequently flying the F-15 aircraft, served to reduce to almost vanishing, individual differences in general and specific psychomotor ability. This would account for their lack of validity for this criterion.

In contrast to the findings of Tett, et al. (1991), Dependability failed to be a significant predictor of the criterion. The reasons for this failure cannot be found in these data.

The implications of the study are straightforward. The first implication is based on the finding that a greater number of F-15 flying hours was related to higher ratings of SA. The more hours pilots are permitted to spend in the F-15 cockpit, the better their SA can be predicted to be. The second implication is related to personnel selection. The current U. S. Air Force pilot candidate selection test battery (AFOQT and BAT) contains measures of the construct found to be predictive of SA; general cognitive ability. Future revisions of pilot selection instruments should retain measures of general cognitive ability.

There are several issues to be addressed in the measurement of cognitive ability. It is necessary to improve the accuracy and completeness of our measures as found in the AFOQT and BAT. The measurement of cognitive ability using several different contents implies that different test level traits may be used. These are often referred to as first-order factors or constructs. Many are familiar, such as verbal and quantitative and some have emerged more recently from models of cognitive components, such as spatial working memory, verbal working memory, and spatial reasoning.

Although the equivalence of new cognitive components and first-order factors remains to be fully investigated, it may be that new cognitive components offer measurement of cognitive ability with almost no content in the usual sense. That is, the new cognitive component tests frequently do not require previous learning other than the language requirements of the instructions. It is conceivable that

303

problems of adverse impact on minority groups and women might be reduced or avoided if new cognitive tests are added to the pilot selection system.

References

Andre, A.D., Wickens, C.D., Moorman, L., & Boschelli, M.M. (1991). Display formating techniques for improving situational awareness in the aircraft cockpit. *The International Journal of Aviation Psychology, 1*, 205-218.

Carretta, T.R., Perry, D.C., Jr., & Ree, M.J. (in press). Prediction of situational awareness in F-15 pilots.

Carretta, T.R., & Ree, M.J. (in press). Pilot Candidate Selection Method (PCSM): Sources of validity.

Goldberg, L.R. (1990). An alternate "description of personalty": The Big-Five factor structure. *Journal of Personality and Social Psychology, 59*, 1216-1229.

Hartman, B.O., & Secrist, G.E. (1991). Situational awareness is more than exceptional vision. *Aviation, Space, & Environmental Medicine, 62*, 1084-1089.

Houck, M.R., Whitaker, L.A., & Kendall, R.R. (1991).*Behavioral taxonomy for air combat: F-15 defensive counter-air mission* (UDR-TR-91-147). Dayton, OH: University of Dayton Research Institute.

Hunter, J.E. (1983). A causal analysis of cognitive ability, job performance, and supervisory ratings. In F. Landy, S. Zedeck, and J. Cleveland, *Performance measurement and theory*, Hillsdale, NJ: Erlbaum.

Hunter, J.E., & Hunter, R. (1984). Validity and utility of alternate predictors of job performance. *Psychological Bulletin, 96*, 72-98.

Kyllonen, P.C. (1993). Aptitude testing inspired by information processing: A test of the four-sources model. *The Journal of General Psychology, 120*, 375-405.

McHenry, J.J., Hough, L.M., Toquam, J.L., Hanson, M.A., & Ashworth, S. (1990). Project A validity results: The relationship between predictor and criterion domains. *Personnel Psychology, 43*, 335-354.

Ree, M.J., & Carretta, T.R. (1992). *The correlation of cognitive and psychomotor tests* (Al-TP-1992-0037). Brooks AFB, TX: Armstrong Laboratory, Human Resources Directorate, Manpower and Personnel Research Division.

Ree, M.J., & Carretta, T.R. (in press). The central role of *g* in military pilot selection.

Ree, M.J., & Earles, J.A. (1992). Intelligence is the best predictor of job performance. *Current Directions in Psychological Science, 1*, 86-89.

Schmidt, F.L., Hunter, J.E., & Outerbridge, A.N. (1986). Impact of job experience and ability on job knowledge, work sample performance, and supervisory ratings of job performance. *Journal of Applied Psychology, 71*, 432-439.

Tett, R.P., Jackson, D.N., & Rothstein, M. (1991). Personality measures as predictors of job performance: A meta-analytic review. *Personnel Psychology, 44*, 703-742.

Part 12
WORKLOAD

46 Developing a flight workload profile using Continuous Subjective Assessment of Workload (C-SAW)

*S.E. Jensen, Man–Machine Integration Department, Defence
Research Agency, Farnborough, UK*

"Reduction of crew workload" is used increasingly as a rationale for the development of sophisticated and expensive cockpit systems. Such a rationale is futile, unless the precise need for this reduction can be defined, and the desired improvement can be shown to have been achieved as a result of the new development. It follows that quantifying workload is pointless unless the data can be readily applied to the improvement of cockpit design.

To apply the results of workload assessment it is necessary to pin-point accurately the tasks that give rise to workload fluctuations. Established subjective methods, such as the Bedford Scale[1], the Subjective Workload Assessment Technique (SWAT)[2] and NASA's Task Load Index (TLX)[3,] relate well to performance, but only for relatively gross coherent 'chunks' of a flight. Computer modelling techniques based on assigning 'workload' values to tasks in the task timeline, while invaluable for guiding an evolving design for a future system, are inevitably based on some very broad assumptions of operator response, and cannot accommodate the range of operator capability and training. Where a simulation or real flight is to be assessed, conventional subjective techniques are too coarse to guide design detail, and it is pointless to risk the compromises of accuracy inevitable in modelling when real subjects are available to operate the real system.

This paper is an account of a workload assessment technique which has been developed in response to the requirement to create a fine resolution flight workload profile whose fluctuations can be precisely related to the tasks and events which give rise to them.

Background

The specific problem which motivated the development of this technique was the need to assess the demands of single-seat operation of a targeting system originally designed for use by the navigator in a two-seat fast jet. The task requires the pilot to fly a complex attack profile using the head-up display (HUD) symbology overlaid on the head-down display (HDD) of the targeting device. The targeting display has to be monitored throughout flight to ensure that a marker remains on the target. Although the DRA test pilots had established that single-seat operation of the device was possible, they were conscious that they were working at the limits of their attentional resources and were very aware of the safety pilot with them, who, while not assisting them with the task, would not allow them to endanger the aircraft. In order to make a more quantifiable assessment of workload, a method of capturing the momentary workload levels through the attack run was necessary.

What was needed was a method of visualising a detailed time-ordered profile of flight workload that could be read across to an equally precise task timeline. An attack run is very eventful and the established subjective and physiological techniques for workload assessment were quite incapable of giving the resolution needed.

Developing Continuous Subjective Assessment of Workload - C-SAW

Taking workload ratings during an attack run would be impossible, even dangerous, so the technique was based on a review of an in-cockpit film of the attack run. Preliminary studies comparing in-flight commentary of subjective workload with the post-flight C-SAW suggested that the aircrew could recapitulate their subjective experience of workload quite consistently, provided the film was viewed immediately after they had landed. While the film played at normal speed, aircrew pressed one of ten keys on a keypad, corresponding to the descriptors of the Bedford Scale[1] in response to a prompt from a computer. The Bedford Scale[1] was used here, as it was necessary to use a unidimensional scale; the Bedford addresses workload specifically, whereas the Modified Cooper-Harper Scale[4] is directed more at errors and the interface design. C-SAW can be used with any uni-dimensional rating scale (not necessarily just workload) with a range from 0 to 100.

Surprisingly, initial 'pilot' runs of the technique showed that the subjects could respond reliably to prompts as frequent as every 3 seconds, provided they were not required to maintain this rate for too long. An attack run is normally complete within two minutes, and the aircrew have little difficulty in maintaining the 3-second input rate for this time. The software collects the data in the form of ASCII text files which can be read into any suitable software package and printed as a bar-chart or graph against the timeline.

Before the experimental flight, a theoretical task timeline is established by consultation with the aircrew, and for each individual flight an accurate timeline is calculated from the video film. The workload ratings and individual task timelines are then combined to provide a flight workload profile and task description with a common timeline.

The data in Figures 1 and 2 are illustrations of C-SAW output from two attack runs on the same target using the same attack profile, but with the system operating fully in Figure 1 and in a degraded but still operationally effective mode in Figure 2. The C-SAW output can be seen clearly to respond to the different workload levels of the two conditions, and the associated on-going tasks can be read off from the task timeline below. NASA Task Load Index (TLX)[3] overall rating and the conventional Bedford Scale ratings for the attack runs are also shown on the C-SAW chart.

C-SAW has been used in flight trials and in simulations, and has achieved a very high face validity. Formal validation is planned in the relatively controllable simulation environment, both with full fidelity simulation and in a multi-workstation computer-based tactical simulator. The criteria used will be test/retest consistency; both for individual subjects and for differences between subjects. Comparison with established techniques is difficult, as these give only an overall rating for the whole time-period, rather than a time-ordered profile, as is illustrated in the two Figures shown below.

Low Level Toss Profile -- Full Operational Mode

Continuous Subjective Analysis of Workload
(C.S.A.W.)

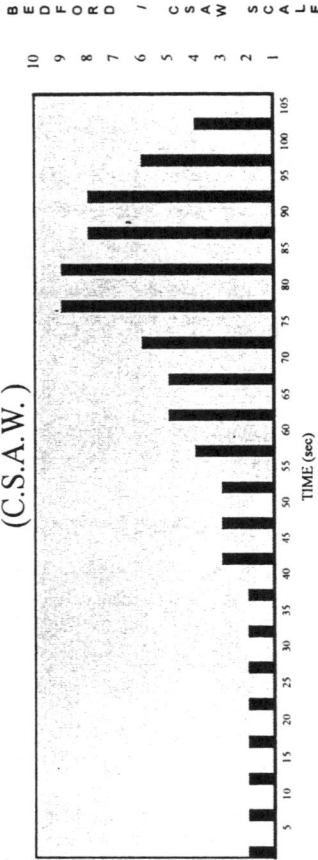

10	BEDFORD
9	
8	/
7	
6	
5	CSAW
4	
3	
2	SCALE
1	

TIME (sec)

Task Timeline

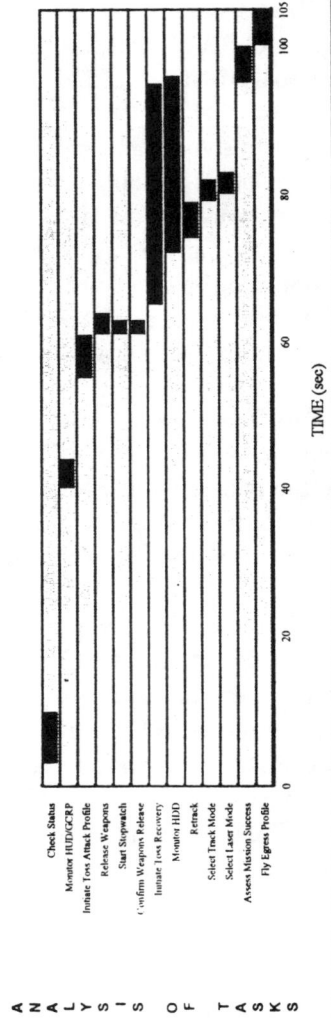

ANALYSIS OF TASKS

- Check Status
- Monitor HUD/GCRP
- Initiate Toss Attack Profile
- Release Weapons
- Start Stopwatch
- Confirm Weapon Release
- Initiate Toss Recovery
- Monitor HUD
- Retrack
- Select Track Mode
- Select Laser Mode
- Assess Mission Success
- Fly Egress Profile

TIME (sec)

310

Low Level Toss Profile – Degraded Mode

Continuous Subjective Analysis of Workload
(C.S.A.W.)

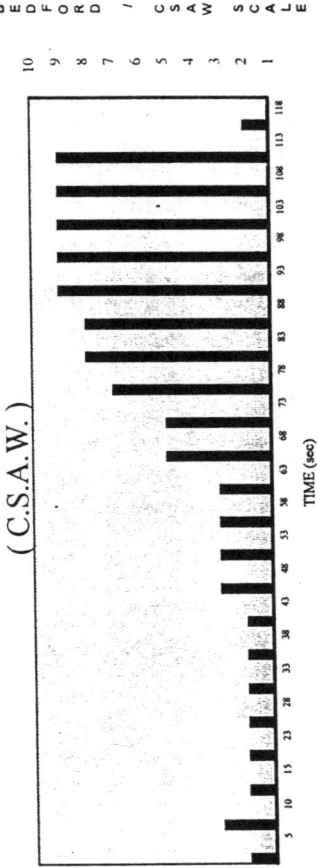

Task Timeline

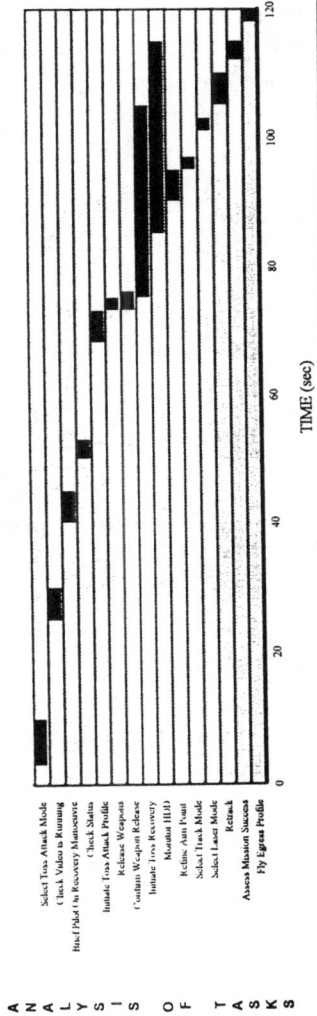

311

The C-SAW technique can also be used in a freeze-frame mode for a very detailed investigation, perhaps when a particular display or manoeuvre is being studied, or where an area of interest identified by the initial 'real-time' assessment needs to be studied in greater detail.

Further work will extend the C-SAW method to the investigation of particular types of workload, such as auditory/verbal workload when communications are being investigated, or aspects of cognitive workload such as short-term memory during complex decision-making. Other areas where it is hoped to assess C-SAW's potential include over- and underload in civil aviation and workload in non-aviation environments, such as process control.

References

Hart, S, Staveland, L E, (1988) 'Development of the NASA TLX: Results of empirical and theoretical research' in Hancock, P A, Meshkati, N, (eds) *Human Mental Workload* North Holland, Amsterdam.

Reid, G.B., Shingledecker, C.A., Eggemeier, T.F. (1981) *Application of conjoint measurement to workload scale development.* Proceedings of the Human Fcators Society 25th Annual Meeting.

Roscoe, A. H., (1987) 'Introduction to practical assessment of pilot workload' in Muir, H.C., Roscoe, A.H., (eds) *Proceedings from the Symposium on the Practical Assessment of Pilot Workload* AGARD Paris.

Wierwille, W. W., Casali, J. G. (1983) *A Validated Rating Scale for Global Mental Workload Measurement Applications.* Proceedings of the Human Factors Society 27th Annual Meeting.

47 Mental workload and performance in combat aircraft: systems evaluation

Erland Svensson and Maud Angelborg-Thanderz
National Defence Research Establishment, Stockholm, Sweden

Pilots in modern military flight systems have to process a considerable amount of complex information. The information and decision making processes have become more and more demanding and the risk for mental overload has increased. Procedures for measuring mental workload (MWL) have become an indispensable prerequisite for analyzing specific missions and evaluating the need of decision support systems.

There is no simple relationship between pilot emotional state and the characteristics of a mission. The same mission can be a threat or a challenge. Positive expectations of being able to handle a situation are helpful and negative expectations may reduce tolerance of frustration. Stress of any kind can make a threat out of a challenge. Even one's mood can tip the scale (Svensson, Angelborg-Thanderz & Sjöberg, 1993a).

Expectations are important determinants of current mood (Sjöberg, 1989). Perceived risk would therefore be an important aspect of a mission, affecting the pilots' mood and hence also their performance (Angelborg-Thanderz & Svensson, 1992).

Great efforts have been spent to define and measure MWL (Williges & Wierwille, 1979; Lysaght, Hill, Dick, Plamondon, Linton, Wierwille,

Zaklad, Bittner & Wherry, 1989). One conclusion is that MWL is a multi-faceted concept, hard to define and capture. Consequently, the concept cannot be measured by one single measure (Gopher & Donchin, 1986).

The relation between MWL and performance is affected by the pilots' capability to cope with the demands of the mission. Gopher and Donchin (1986) define MWL as "the difference between the capacities of the information processing system that are required for task performance to satisfy performance expectations and the capacity available at any given time" (p. 41-3). MWL is accompanied by a psychological and physiological cost caused by the performance of the task.

The techniques for measuring MWL can be divided into three main categories: subjective measures, performance or task measures, and physiological measures. Unfortunately, none of the assessment techniques has a transparent theoretical basis.

Modified versions of the Cooper-Harper Scale (Roscoe, 1987; Wierwille & Connor, 1983) are frequently used in the aircraft industry as a measure of MWL. Subjective Workload Assessment Technique (SWAT) (Reid & Nygren, 1988) and NASA Task Load Index (NASA-TLX) (Hart & Staveland, 1988) exemplify frequently used multidimensional rating scales.

Data from two studies have been utilized in the development of a workload index for military pilots. A first study concerned complementary attack training and a second intermittent fighter flight training. From a psychological model based on data from study 1, a set of variables reflecting different aspects of MWL has been selected. The validity of the workload indicators has been tested on data from study 2 (Svensson, Angelborg-Thanderz, Sjöberg & Gillberg, 1988; Svensson et al., 1993a). Our basic thrust differs from more traditional measurement validation studies in the application of directed causal structure models. Methodology for this type of approach is provided by 2nd generation multivariate statistics (LISREL; Jöreskog & Sörbom, 1984). Performance is modelled as a result of MWL, and workload is construed as a consequence of mission characteristics.

A factor analysis of the manifest variables from Study 1 was the starting point for the causal modelling. The model has its starting point in a challenge factor (i.e. difficulty and risk) and its terminal point in performance. Two intervening processes have been named problem-solving and emotion-coping, respectively. Both of these processes are affected by the challenge factor. The final model is presented in Figure 1.

314

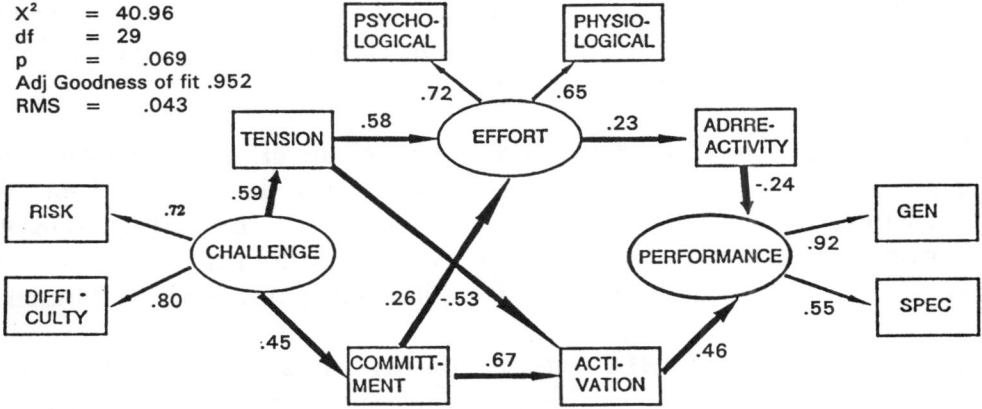

Figure 1 Structural LISREL model of the relationships between the manifest variables. Thin arrows = Factor loadings; Broad arrows = Factor effects. All factor loadings and effects are significant (p < .05).

The problem solving process is characterized by commitment and activation. Increased activation is mediated by increased commitment. Activation and commitment indicate psychological "energy mobilization" which promotes efficient problem solving, decision making, and direct action and, accordingly, has a positive value. The problem solving process is positively related to the performance factor in the model.

The emotion coping process is characterized by tension, effort, and adrenaline reactivity. Increased challenge results in increased tension which, in turn, (a) increases effort and (b) decreases activation. The emotion coping process is negatively related to performance.

The variables included in or directly affected by the emotion coping process i.e. tension, psychological and physiological effort, adrenaline reactivity, and activation (inverted) constitute the markers of our MWL index. According to the proposed set of markers, high workload is characterized by increased tension (mental stress), increased psychological and physiological effort, psychophysiological energy mobilization and, sooner or later, fatigue.

The model from study 1 was validated in study 2 even if a somewhat different internal structure was found. In the second study the effects of intermittent flight training on MWL and performance were studied. We found that our workload index changed significantly as a function of a training program including 25 missions (r = -.43; p < .01). The correlation was the same even when the effects of challenge and training were controlled for. Our conclusion was that the significant reduction in MWL was a genuine effect of the training program.

The purpose of the third study was to analyze the effects of information complexity on MWL and performance, and to test the practical applicability of different measures of MWL. This study was the starting point of our ongoing research on evaluation of MWL and pilot performance in the systems development of the multirole military aircraft JAS 39, the Gripen.

The complexity of the HDD information was varied as a function of the tactical situation in 72 simulated low level - high speed missions. Flight data were registered. The pilots' eye movements were video taped and their psychophysiological activation (HR) was saved on discs. During and after missions the pilots rated MWL according to 3 scales [BedFord Rating Scale (BFRS), SWAT, NASA-TLX] and psychological variables.

It was found that even a moderate complexity of information interfered with the flight task. The pilots' altitude and variation in altitude were increased, and their corrections of altitude errors were delayed as a function of information load. Durations and frequencies of eye fixations (Head Up - Head Down) changed as a function of information load and the conditions for flying low level-high speed with precision got worse when the information load HD increased. When the information load on the Tactical Situation Display (TSD) exceeded a certain level the pilots could no longer integrate information handling and flying and they succeded with none.

The psychological indices of MWL covaried with our indices of difficulty, mental effort, time stress, capacity, saturation, and complexity TSD. Based upon the covariations between the indices, MWL measures, objective registrations, and instructor ratings, a structural equation model how different factors affect and are affected by MWL has been developed. The model has its starting point in the task related indices complexity TSD and difficulty, and its terminal point in different aspects of the pilots'performance. The MWL forms an intervening process. The perceived difficulty increased as a function of the complexity of the tactical information. The increasing difficulty resulted in decreased performance and increased MWL. An increased MWL caused changes in the objective registrations. The number of eye fixations HD and the variation in speed increased, and the precision of information handling decreased as a function of MWL. Changes in the objective registrations affected the instructor ratings of the pilots' performance. An increased number of critical eye fixations HD caused

316

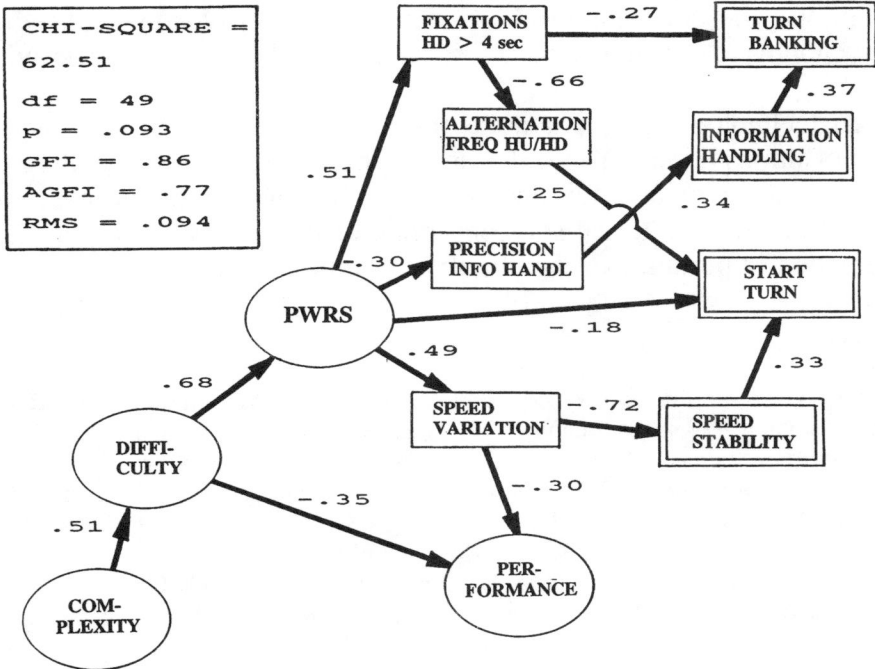

Figure 2 Structural causal model (LISREL) of the relationships between different indices, MWL measures (PWRS), objective registrations, and instructor ratings. Pilot ratings are marked by ellipses, objektive registrations by simple rectangles, and instructor ratings by double rectangles. Effects > .24 are significant (p < .05).

a decreased performance rating of turn banking. The performance rating of information handling decreased as a function of the difference between presented and reported number of objects on TSD. The ratings of speed stability decreased as a function of an increased speed variation, and an increased speed variation caused a decreased pilot rating of his performance. To sum up, MWL was sensitive to changes in the difficulty levels of the task, and MWL predicted objective as well as subjective measures of pilot performance. The model is presented in figure 2.

Of specific interest was the significant covariation (r = -.35 to -.44) between mood and MWL. The more alert, relaxed, and confident the pilot was before a run, the better was his ability to cope with the load of it.

Heart rate correlated positively with MWL and complexity of mission, and it covaried with variations in information complexity over the sortie for those pilots who performed well. Compared to the psychological indices the psychophysiological index disclosed a restricted differential sensitivity.

Modern technology makes it quite feasible to provide the pilot with a wealth of information. Surely, the limit of performance is set by the limitations of the human operator and not by technological possibilities. In this situation we feel that a realistic conception of the human limitations is of fundamental importance in the design of combat aircraft.

References

Angelborg-Thanderz, M. & Svensson, E. (1992), 'Good-value military flight training with reasonable risks', in Julander, C.-R. & Wahlund, R. (eds), *Perspectives in Economic Psychology,* Stockholm School of Economics, Stockholm.

Gopher, D. & Donchin, E. (1986), 'Workload - an examination of the concept', in Boff, K.R., Kaufman, L. & Thomas, J.P. (eds), *Handbook of Perception and Human Performance. Vol.II,* New York, John Wiley & Sons.

Hart, S.G. & Staveland, L.E. (1988), 'Development of NASA-TLX (Task Load Index): results of empirical and theoretical research', in Hancock, P.A. & Meshkati, N. (eds), *Human Mental Workload,* Amsterdam, Elsevier.

Jöreskog, K.G. & Sörbom, D. (1984), *LISREL VI: Analysis of Linear Structural Relationships by Maximum Likelihood, Instrumental Variables, and Least Squares Methods,* Uppsala, Sweden, University of Uppsala, Department of Statistics.

Lysaght, R.J., Hill, S.G., Dick, A.O., Plamondon, B.D., Linton, P.M., Wierwille, W.W., Zaklad, A.L., Bittner, Jr, A.C. & Wherry, R.J. (1989), *Operator Workload: Comprehensive Review and Evaluation of Operator Workload Methodologies,* U.S. Army Research Institute, Fort Bliss, TX, Technical Report No. 851.

Reid, G.B. & Nygren, T.E. (1988), 'The subjective workload assessment technique: a scaling procedure for measuring mental workload', in Hancock, P.A., Meshkati, N. (eds), *Human Mental Workload,* Amsterdam, Elsevier.

Roscoe, A.H. (ed), *The Practical Assessment of Pilot Workload,* Neuilly-sur-Seine, AGARD/NATO, AG 282.

Sjöberg, L. (1989), 'Mood and expectation', in Bennett, A.F. & McConkey, K.M. (eds), *Cognition in individual and social contexts,* Amsterdam, Elsevier.

Svensson, E., Angelborg-Thanderz, M., Sjöberg, L. & Gillberg, M. (1988), 'Military flight experience and sympatho-adrenal activity', *Aviat. Space Environ. Med.,* **59,** 411-16.

Svensson, E., Angelborg-Thanderz, M. & Sjöberg, L. (1993a), 'Mission challenge, mental workload and performance in military aviation', *Aviat. Space Environ. Med.,* **64,** 985-91.

Svensson, E., Angelborg-Thanderz, M., Sjöberg, L. & Olsson, S. (1993b), *Information Overflow? Mental Workload and Performance in Combat aircraft,* Workload Assessment and Aviation Safety, Royal Aeronautical Society, London.

Wierwille, W.W. & Connor, S.A. (1983), 'Evaluation of 20 workload measures using a psychomotor task in a moving-base aircraft simulator', *Human Factors,* **25,** 1-16.

Williges, R.C. & Wierwille, W.W. (1979), 'Behavioral measures of aircrew mental workload, *Human Factors,* **21,** 549-74.

318

48 AWAS (Aircrew Workload Assessment System): issues of theory, implementation and validation

A.K. Davies, A. Tomoszek, M.R. Hicks, J. White
Sowerby Research Centre, British Aerospace PLC

The development of increasingly complicated hardware and software systems demands the optimisation of the interaction between operator and machine. A poorly designed cockpit system can lead to aircrew overload - during periods of high workload, some tasks may be performed less efficiently, or with errors, and some tasks may be shed altogether. Analysis of the system, at all design stages, to determine the expected workload of the aircrew during different phases of flight is therefore advantageous. Techniques to determine human performance and workload are now becoming available, and with further development may prove to be extremely valuable in the design of cockpit systems.

Mental workload is difficult to define, even though human beings have an intuitive understanding of workload and can recognise when their workload is too high. Workload is a subjective experience, mediated by both internal and external factors, but it does have a direct and measurable effect on human performance. Mental workload cannot be measured in absolute units, but it is possible to compare the relative workload produced by different situations, and it is therefore possible to define sets of tasks, or different sequences of tasks, which produce lower workloads than other combinations. The ability to identify "hotspots" in various phases of flight may allow system design to be adapted to reduce workload - tasks can be re-assigned between aircrew, or can be taken over by the flight computer during periods of high aircrew workload. Efficient interaction between the aircrew and cockpit systems will also reduce the likelihood of errors, and thus lead to enhanced safety. Accurate methods for predicting workload would therefore seem to have great potential. This paper will discuss current theories of workload prediction, and present work designed to provide validation data for a variant of one theoretical implementation (Wickens et al., 1988) that has been under development at British Aerospace.

The number of publications concerned with workload has increased in recent years, and a number of techniques have shown promise. One approach is based on "time-line" analysis - a second by second analysis of activity over time. A number of computer tools have been created based on this approach (eg.WINDEX (North and Riley, 1988); AWAS (Hicks, 1993)).

Workload estimates are derived from time-line information in a variety of ways. One approach (multiple resource theory) assumes that humans have several different information processing resources that can be tapped simultaneously. This assumption seems reasonable; humans receive information in several forms (visual, auditory, motor etc.), and may respond in a variety of manners. Some combinations of tasks are easy to perform simultaneously, perhaps because they tap separate information processing channels which process information in parallel. Other combinations of tasks are not so easy to perform simultaneously, perhaps because they require the same information processing channels and therefore can only deal with the information in a serial fashion. If human information processing is achieved via some limited number of processing resources or channels, then the possible channels must be defined, and tasks described in terms of the level of activity in these channels. Different levels of activity in a particular channel may be possible - some tasks may make significant demands on a particular channel, whereas other tasks might only use up a fraction of that channel's information processing capacity.

The next stage in the development of this particular approach to modelling human information processing is to consider what happens when an operator is required to perform two or more tasks simultaneously. The multiple resource theory of information processing proposes that workload may be assessed by consideration of the conflicts occurring between the processing channels required for each task. Certain task combinations may be acceptable; others may lead to high workload. A generalised form of workload equation, on which a number of the algorithms described in the literature have been based (eg. Wickens, 1988), may be defined as:

$$workload = \Sigma \begin{matrix} workload\ produced \\ by\ individual\ tasks \end{matrix} + \Sigma \begin{matrix} workload\ produced\ by\ conflicts\ between \\ and\ within\ resource\ channels \end{matrix}$$

Application and validation of workload theories

This paper describes a study designed to obtain data to validate and develop workload models. These data can be applied to the assessment of the

320

reliability of any workload model, although in this case the results are tested against AWAS (Aircrew Workload Assessment System), a software tool created at British Aerospace. AWAS is based on a time-line analysis approach and includes a module for workload predictions based on the multiple resource theory (Wickens et al, 1988).

For the purposes of this study the AWAS model requires three distinct sets of inputs. Firstly, a time line analysis is made of the tasks the pilot performs in the form of a second by second description of the flight. Secondly, the demands likely to be made of specific information processing channels by each task are identified. Thirdly, the effect on workload arising from conflicts caused by simultaneous demands on pairs of information processing channels has to be defined. Given this information the model makes an assessment of the instantaneous workload sustained by the pilot at any moment during the flight (for example, see Fig. 1).

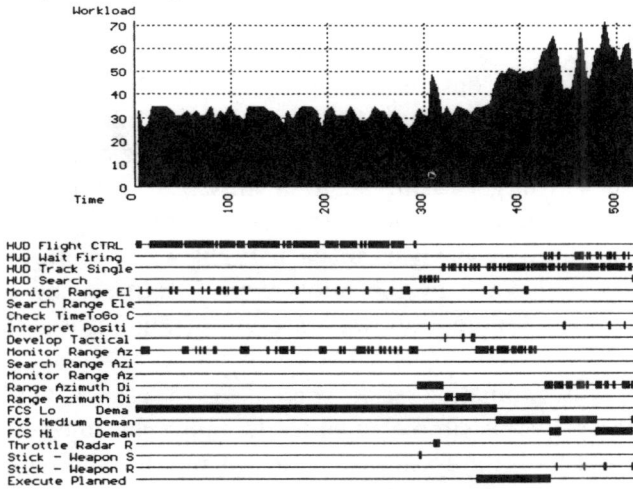

Figure 1 Sample time line information for a simulated flight. Activities of 21 tasks are depicted along with the predicted workload.

The validation of any mathematical model of human performance requires experimental data with which to assess the model's performance. All mathematical models should be tested against experimental data, but the assessment of workload models is complicated by the fact that direct measurements of workload are impossible and instead must be inferred by measuring some variable with which workload is known to correlate.

Certain physiological measurements have been shown to depend on mental activity. Recordings may be made of cardiac activity, blink rate, pupil diameter, EEG (electroencephalogram), and ERP (event-related brain potentials) and all correlate to some degree with mental demands.

Alternatively workload may be rated using performance based assessment techniques. Here it is assumed that the proficiency of the operator in performing some aspect of the task directly reflects the workload. For instance, the workload imposed by different instrument configurations could be assessed by reference to the deviation of an aircraft from a glide slope. If it is not convenient to analyse workload in terms of operator primary task performance, then it may be necessary to define workload in terms of the operator's proficiency on a secondary task. Many authors have conducted encouraging validations of mental workload models (eg Wickens et al., 1988; Sarno and Wickens, 1991) and the literature shows that many different workload models exhibit comparable predictive powers. This paper describes work undertaken to validate two variants of a workload model - one based on multiple resource theory and one based on a simpler task-counting approach.

Method

The present study was designed to provide workload estimates and task analysis for a fast jet engagement. Two experienced pilots flew a total of five missions in a Sea Harrier Simulator, in each case being asked to approach and engage a number of targets using air to air missiles. Flight durations ranged from 3.5 to 9 minutes. Video recordings were made of each flight. Subsequently a time line analysis of each flight was generated whereby the individual tasks performed by the pilots were identified and their activities recorded over time. Figure 1 shows a section of this information for one of the flights.

An auditory discrimination task was used to assess workload. This secondary task was kept simple in order to minimise interference with the primary task, and took the form of a series of rapidly alternating tones. On average, every 14 seconds one of the tones was omitted (ie two repetitions of one of the tones rather than the normal alternating pairs) and the pilot responded verbally to this event. Secondary task performance was recorded throughout each flight.

Results

If the workload scores predicted by AWAS for each flight are representative of the actual demands placed on each pilot, then a good degree of correlation should exist between the probability of a secondary task error and the predicted workload score at the instant that the secondary task was presented. This is confirmed by reference to Figure 2 which shows a clear correlation (r=0.904) between probability of secondary task error and AWAS workload prediction. This result indicates that the AWAS implementation of the multiple resource theory is a good predictor of workload.

The multiple resource theory depends on the interaction of the data from the conflict and demand databases, implemented mathematically by matrix multiplication and addition. Replacing all of the figures in these databases by unity "short-circuits" the multiple resource approach - all tasks are given equal weighting and workload predictions are directly based on the number of tasks which the pilot is performing simultaneously. Figure 3 shows the relationship between these new workload predictions and the probability of a error on the secondary task. A correlation is again evident (r=0.983).

Discussion

These results suggest that the currently implemented multiple resource algorithm may be no more sensitive than the considerably less sophisticated approach of basing workload assessments on the number of

Figure 2 Correlation of probability of error on the secondary task with the predicted workload according to multiple resource theory.

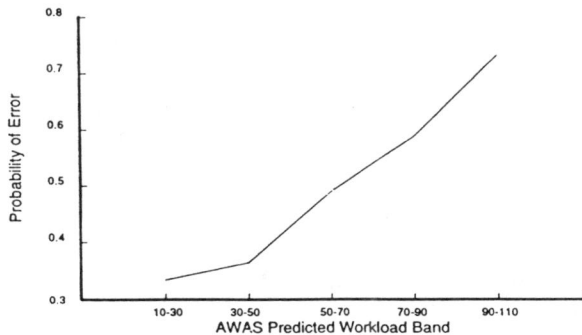

simultaneous tasks which a pilot performs. Indeed the correlation indicated for the simple task-counting approach was higher than obtained using the sophisticated multiple resource technique. These findings conflict with the evidence presented by Wickens (1988) who suggested that performance was better predicted by resource conflict models than by simpler time-line analysis only models. A number of points must be stressed, however. Firstly, the results may reflect shortcomings in the validation process itself. The choice of secondary task was such that the total number of trials was limited and it might be preferable if a greater sample were available for analysis. Thus, a performance metric based directly on the primary flying

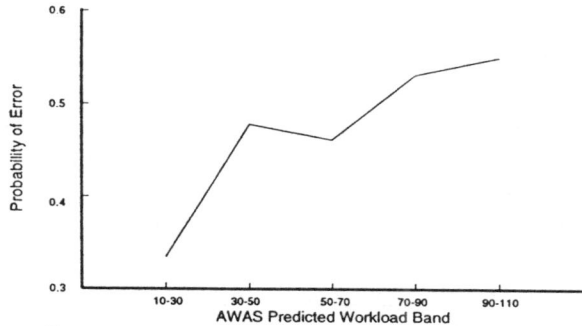

Figure 3 Correlation of probability of error on the secondary task with the predicted workload based on current number of active tasks.

task might have been more reliable. Secondly, it is likely that an improved version of the task resource database could be created which would further increase the reliability of predictions. It should be stressed, however, that both implementations of the workload algorithm provided reliable predictions of secondary task errors - additional work will be conducted to more fully assess the relative merits of the two approaches.

One difficulty of the multiple resource approach to workload prediction is the complexity of channel definition, that is the decomposition of each task into a pattern of channel activity and the allocation of conflict values between channels. Often these values are assigned by "subject matter experts", rather than being derived from experiment. In this respect simpler approaches may prove more appealing. The drawback of such simple models is that they cannot purport to model actual cognitive processes but will merely provide convenient curve-fits to experimental data. One major attraction of the multiple resource approach is that it attempts to represent underlying psychological processes and with further development may form the basis of a more rigorous approach to workload modelling.

Finally, this study stresses the importance of validating performance models not in isolation but in conjunction with other alternative models. The validation of one particular approach should only be considered convincing if it can be shown that alternative methods are not more reliable. The gathering of validation data should remain a high priority in workload modelling. The establishment of a comprehensive database against which models may be tested will facilitate the testing and comparison of different models. Once established such a database should allow the full potential of mental workload models to be realised.

References

Hicks, M.R. (1993), AWAS analytical workload modelling tool - research and development programme, Presented RAS Symposium "Workload Assessment and Aviation Safety", London.

North, R. & Riley, V. (1988), WINDEX: A predictive model of operator workload. In G. MacMillan (Ed.), *Human Performance Models*, NATO AGARD Symposium, Fl.

Sarno, K. and Wickens, C.D.(1991), The role of multiple resources in predicting time-sharing efficiency : an evaluation of three workload models in a multiple task setting,NASA Ames Technical Report, Contract NASA NCC 2-632.

Wickens, C.D., Harwood, K., Segal, L., Tkalcevic, I., Sherman, B., (1988), TASKILLAN : A simulation to predict the validity of multiple resource models of aviation workload, Proceedings of the human factors society, 32nd annual meeting.